Contents

Contractor's Claims
an architect's guide

Contractor's Claims
an architect's guide

David Chappell

The Architectural Press
London

First published in 1984 by The Architectural Press Ltd,
9 Queen Anne's Gate, London SW1H 9BY

© David M. Chappell 1984

British Library Cataloguing in Publication Data

Chappell, David
   Contractor's claims
   1. Building—Contracts and specifications—
   Great Britain
   I. Title
   344.103'78624   KD1641

ISBN 0–85139–778–6

Acknowledgements
Extracts from JCT80 and the Agreement for Minor Building
Works appear with the permission of the copyright holders
RIBA Publications Ltd

Extracts from the ACA Form of Building Agreement 1982
appear with the permission of the copyright holders the
Association of Consultant Architects

Extracts from the GC/Works/1 contract are Crown copyright
and are reproduced with the permission of the Controller of
Her Majesty's Stationery Office.

Typeset in Great Britain by Fakenham Photosetting Ltd,
Fakenham, Norfolk
Printed in Great Britain by Biddles Ltd., Guildford, Surrey

Introduction

Contractor's claims reach the architect in many forms, for varying amounts and periods of time and supported by a wide range of evidence. A great deal of the architect's time is taken up by consideration and determination of such claims.

Little positive guidance has been given to the architect about the actual methods to be adopted in deciding claims. This is partly because no two claims are ever exactly the same and partly because the process is so inherently complicated. It is sometimes held that it is impossible to use any particular system to determine claims. This view must be wrong, otherwise no claims would be concluded or, if concluded, would be the subject of arbitration in every case.

It would be arrogant to suggest that this book will solve all the architect's problems when confronted with a claim. It is intended to give practical guidance so that architects know how to react when they receive claims, some of the methods by which they can arrive at decisions and what to do about their decisions.

Chapters are included dealing with the contractor's duties, the value of various types of evidence, liquidated damages and penalties, certificates and employer's decisions. All of these topics are discussed because they are inextricably bound up with the main subject of the book—deciding claims.

This is not intended as a legal textbook. Although the law enters into the consideration of claims, discussion of legalities is kept to a minimum. It is a book for architects. However, contractors will also find it useful as an indication of the way their claims are likely to be treated and as a guide to the best way to present them to obtain speedy resolution.

The contracts to which references are made are the JCT80 SF with Quantities, the ACA82 Building Agreement, the GC/Works/1 and the JCT Agreement for Minor Building Works. The claims provisions in the various contracts are often extremely complex. This book attempts to simplify them. Quotes from the original documents are kept to a minimum and provisions have been rephrased in the interests of clarity. It is assumed that the reader will have the appropriate documents to hand.

At the time of writing, a new edition (1984) of the ACA form is in course of preparation. It is anticipated, however, that claims will be current under the 1982 form for some time to come. The 1984 edition will be taken into account in future editions of this book.

The text is arranged to make it as easy as possible to follow the requirements of a particular contract by using alternative chapters as appropriate. A glance at the table of contents will make the system clear. Simple flow charts for all the main procedures are included for quick reference and sample letters are provided to deal with a number of common situations. The letters should be amended as necessary to suit the precise situation.

For convenience, the masculine gender has been used throughout and 'he' may be taken to mean 'she', 'his' to mean 'hers' etc.

Every letter should have a heading stating the job title. Headings have been omitted in the examples illustrated.

My thanks must go to the contractors I have encountered over the years, who perplexed me with their ingenuity, and my colleagues who rescued me with good advice.

1 What is a claim?

1.1 Introduction

A claim is a demand that one's rights be satisfied.

The word 'claim' is very emotive in the construction industry. Contractors are often referred to as being 'claims conscious'. This type of contractor is generally disliked by the architect and the employer. The use of the phrase is unreasonably attached to contractors who make a habit of presenting claims during a contract. It does not usually have any bearing on whether the claims are justified or not. Very often the contractor will not make a claim because he does not have the knowledge or time to present it properly. Architects dislike dealing with claims because:

- It occupies time and energy that could be devoted to work elsewhere.
- It implies increased costs, which will displease the employer.
- It may imply some lack of care on the part of the architect, which is embarrassing and could lead to some action by the employer.
- Many architects do not know how to deal with claims properly.

From the contractor's point of view, he wishes his claim to be dealt with speedily and accurately so that he is sure of his position as far as time and money are concerned.

1.2 Types of contract

The four forms of contract which will be considered are:

- JCT80
- ACA82
- GC/Works/1
- Agreement for Minor Building Works.

JCT80 and the Agreement for Minor Building Works are negotiated documents, agreed by all sides of the construction industry. In contrast, ACA82 and GC/Works/1 are 'written standard terms of business' (*Unfair Contract Terms Act*, 1977) and in any proceedings before the courts these forms may be construed against the maker (the employer).

There are some differences in terminology between the four contracts, as shown in table 1 below.

Table 1   Terminology of contracts

JCT80	ACA82	GC/Works/1	Agreement for Minor Works
Relevant event	Circumstance	Circumstance	
Matter	Event		Not applicable
Loss and/or expense	Damage, loss and/or expense	Expense	Not applicable
Practical completion	Taking-over	Completion	Practical completion
Employer	Employer	Authority	Employer

Where it is necessary to use one term to cover all four contracts, the JCT80 terms have been adopted. Otherwise, each contract is described using its own terms.

1.3
Types of claim

There are two basic types of claim:

● Claim for extra time to complete the contract.
● Claim for extra money arising out of the contract.

They may, or may not, be connected. Financial claims can be further divided into:

● *Contractual claims*, which include any claims made by the contractor under the express provisions of the contract—variations and loss and/or expense are examples. This is the only class of financial claim which you are empowered by the contract to determine.
● *Ex-contractual claims*, which are claims made outside the express provisions of the contract and

usually as a result of a breach of its terms, which
may be express or implied.

● *Ex-gratia claims*, which are claims for which
there is no legal basis but for which the contractor
considers he has moral grounds.

Although extension of time and financial claims
are often presented separately, financial claims are
sometimes a mixture of all three sub-divisions. It
is essential for you to be able to differentiate
between them and keep them quite separate. If
you do not, the result is inevitably a drawn out
dispute during which all parties become
estranged—with serious effects upon the contract
as a whole. The greatest single reason for conflict
is not normally the actual substance of the claim,
but the fact that the contractor has presented it in
a confused way. The architect has then attempted
to judge it as it stands and subsequently passed it
on to the quantity surveyor in despair. The
quantity surveyor, if he had any sense, would then
have passed it smartly back to the architect.
Meanwhile, the contractor will have complained
that he has not received the money due to him.

The most common misconception among
architects, quantity surveyors and contractors is
that financial claims, like extensions of time, are to
be considered on the basis of days or weeks. For
example, the contractor submits a claim for
extension of time and in due course the architect
awards an extension by referring to the relevant
events. The contractor then selects those relevant
events which are similar to the matters under the
loss and/or expense clause and claims additional
costs based on the appropriate number of weeks
extension. The architect merely checks that the
contractor has referred to the correct matters and
then passes the claim to the quantity surveyor to
agree payment on a £x per week basis. This
scenario must be familiar to all architects and
contractors, whether they practise it or not.

It has been said many times before, but bears
repeating, that this method of evaluating

additional costs is not correct. The extension of time clause and the loss and/or expense clause of a contract should be kept quite separate, as one does not follow from the other. The award of an extension of time implies two things:

● The contractor is relieved from the liability to pay (or to have deducted) liquidated damages for the period of time awarded.
● The employer preserves his right to deduct liquidated damages for any subsequent period.

It does not imply any entitlement to additional costs. On the other hand, it could be perfectly reasonable for the contractor to claim financially even if there has been no extension of time or if he has, in fact, completed before time.

1.4
Extension of time

A clause enabling you to award an extension of time is specifically for the benefit of the employer. If such a clause was not included in the contract and some event under the control of the employer (such as non-provision of instructions) prevented completion by the completion date in the contract, time would become 'at large'. In other words, there would no longer be a fixed date for completion, merely a reasonable period of time (whatever that may mean!).

Apart from the fact that the employer would not be sure when the contract would be completed, he would not be able to enforce any claim for liquidated damages on the contractor. If some other event, such as a major strike, outside the contractor's control rendered the contract impossible to perform by the completion date, the result could be similarly disastrous. It is, therefore, of great importance that, in issuing your award, you must:

● grant a fair and reasonable period; and
● do so within a reasonable time of receiving the contractor's claim.

The main forms of contract have specific provisions.

1.4.1
JCT80

Clause 25 sets out the provisions for extension of time. The respective duties of contractor and architect are laid down together with a detailed list of relevant events, which may give rise to a claim, and a time scale for the issue of your award.

Clause 24 specifies that the contractor must pay, or the employer may deduct from moneys due, liquidated damages calculated at the rate stated in the Appendix for the period between the contractual completion date (or extended date) and the date of practical completion. If the contractor does not complete by the contractual completion date, you are obliged to issue a certificate to that effect and the employer's right to recover liquidated damages depends upon the issue of that certificate.

1.4.2
ACA82

Extensions of time are covered by clause 11. It has two sets of alternatives:

● The employer may deduct liquidated damages or he may deduct unliquidated damages (clause 11.3, alternatives 1 and 2).
● Extensions may be granted only as a result of some act etc of the employer or architect, or as a result of a broader range of events, which are still limited in comparison with the JCT80 form, however (clause 11.5, alternatives 1 and 2).

In each case, one of the alternatives must be deleted.

You would be wise to consider carefully before advising the employer to delete the provision for liquidated damages since, as discussed in chapter 8, they provide a simple and relatively trouble-free method whereby the employer can recover pre-estimated loss. Alternative 2 appears likely to be difficult to operate in practice because unliquidated damages must be proved and litigation could well result. Clause 11.5

(alternative 2), which allows a broader range of claimable events, is probably preferable, if only on the grounds that the contractor will increase the amount of his tender if it is deleted.

If the works are not fit and ready for taking-over on the date or dates stated in the time schedule, you are obliged to issue a certificate to that effect. Deduction of damages (liquidated or unliquidated) is conditional upon your certificate.

Note that *liquidated damages* are expressly stated to be deducted by the architect from the amounts payable on any certificate, if the employer so wishes, or the employer may recover them as a debt. In practice, this means that you should produce the certificate in the normal way, with an additional calculation below to show the amount deducted. (A special form is produced by the ACA for this purpose.) The employer should then pay the amount on the certificate, which will have had the amount of liquidated damages already subtracted. At the time you issue your certificate that the works are not fit and ready (see letter 1), you must obtain from your employer specific written instructions to deduct.

Unliquidated damages are left entirely for the employer to deduct if he so wishes.

1.4.3
GC/Works/1

Provisions for extension of time are contained in clause 28. The respective duties of contractor and authority are laid down. This form makes the authority responsible for deciding claims for extension of time. If you are acting as the authority's architect, the job will fall upon you. The change of emphasis is more apparent than real. The JCT80 Local Authorities form cites the architect as the person responsible for estimating extensions of time, but very often the architect is employed within the local authority's own organisation. The point to watch, however, is that the award must be issued from the authority. For the sake of convenience, it will be assumed that you act on behalf of the authority in

1   Letter from architect to employer requesting instructions regarding liquidated damages
This letter is suitable only for use with ACA82

Dear Sir,

I enclose my certificate in accordance with clause 11.2 of the contract.

You may now take steps to recover liquidated damages at the rate stated in the Time Schedule or I can arrange to deduct the appropriate amount from the next certificate, due to be issued on or about the [*insert date*].

If you decide that the latter course is most convenient, I should be pleased to receive your written instructions as soon as possible.

Yours faithfully,

estimating extensions of time. A detailed list of events (referred to as circumstances) which may give rise to a claim for extension is set out in the clause.

Liquidated damages (clause 29) are payable by the contractor when:

● the date for completion has passed and
● the works are not completed and the site is not cleared and
● the authority has given notice to the contractor that he is entitled to no, or no further, extension of time.

1.4.4
Agreement for Minor Building Works

Clause 2.2 sets out the provisions for extension of time. They are very short. There is no detailed list of events; simply a general reference to 'reasons beyond the control of the contractor'. Although there is no time scale laid down for the issue of your award, you must carry out your duty within a reasonable time.

Clause 2.3 deals with liquidated damages. There is no requirement for you to certify that the works should have been completed by a certain date, but the contractor must pay liquidated damages to the employer at the rate stated in the contract. The employer can, of course, waive his right to such damages if he wishes. There is no provision for the employer to deduct.

The lack of provisions found in the larger JCT80 form is due to the relatively small value and short length of contract expected to be carried out under this form.

1.5
Loss and/or expense

All forms of contract under consideration, except the Agreement for Minor Building Works, make provision in varying degrees for the contractor to claim loss and/or expense. A contractual provision sets out rules and conditions intended to cover the most commonly encountered situations. The benefit of a contractual right is that it removes the necessity (unless the contractor decides otherwise) for a direct application to the employer, which might easily result in proceedings before the courts.

1.5.1
JCT80

The principal clause allowing a claim for loss and/or expense is clause 26. It makes provision for the contractor to claim 'direct loss and/or expense' if 'the regular progress of the Works or any part thereof has been or is likely to be . . . materially affected' due to any of the matters noted in the clause.

The precise meaning of this clause has been the subject of some debate among legal commentators. It is inevitable that an approach that appears to be correct for one case may be thought to be unjust or too generous for another. It is essential, therefore, to decide what the clause means in its raw state and then apply it consistently thereafter.

The word *'direct'* in this context refers to damage that is a direct consequence of one or more of the matters noted in clause 26. In other words, there must be no intervening cause of the damage. For example, if you were in default with regard to clause 26.2.1 (late instructions), the contractor would be entitled to recover his loss incurred due to such default. However, if the contractor was, in fact, prevented from dealing with that part of the work because of his own difficulties, the late instructions would not be the direct cause of the loss and he would not have a valid claim. Undoubtedly, you would be in default, but there would be no damages.

Another way to look at this particular problem would be to consider what the situation would be if the instructions had been provided at the proper time. In the example noted above, the contractor would still have suffered exactly the same loss due to his own fault.

If you issue a late instruction to vary the type of sanitary fittings, following which the contractor places an order that the supplier subsequently fails to meet on the agreed delivery dates, the contractor will be able to claim loss and/or expense due to the late instruction. He will not be able to claim from the employer for loss and/or expense due to the failure of the supplier to meet the delivery dates. Undoubtedly, the failure was related to the instruction (if we assume that the original supplier would have delivered on time) but it was not a direct result of the instruction. Of course, the contractor would have recourse to the supplier for his loss.

To take another example, if—as a result of late instructions—the contractor is delayed on a particular piece of external work which is then delayed further by 'exceptionally adverse weather conditions', it is thought that his loss would be reimbursable to cover the exceptionally adverse weather conditions in regard to that particular piece of work. Exceptionally adverse weather conditions (clause 25.4.2) normally qualify only for an extension of the contract period.

Direct loss and/or expense flows directly from the breach. All losses and expenses that can fairly be said to arise naturally in the ordinary course of things are recoverable. Indirect (or consequential) loss and/or expense extends beyond that to embrace all losses and expenses that occur by reason of some special or unusual circumstance. The phrase *'loss and/or expense'* means simply damages as they are commonly understood. The phrase *'regular progress of the Works'* refers to the progress of the works in an orderly fashion so as to achieve completion by the contractual completion date. It is probably reasonable to say that it refers to the progress planned and envisaged by the contractor when he produced his master programme, provided that his programme can be shown to be realistic (but not otherwise). It does not imply that the overall contract period is necessarily extended. Finally the phrase *'materially affected'* means that the progress must be substantially affected. Small interruptions are, therefore, excluded from consideration.

One other important point about the

contractor's claim is that it must concern a matter for which he cannot recover payment under any other provision of the contract. For example, he can recover his losses due to the *late* issue of an architect's instruction, but he will be reimbursed under clause 13 for the costs involved in any additional work contained in the instruction.

The contractor, therefore, can claim under this clause any loss incurred if:

● One of the matters was the cause of the loss and not merely an accompanying circumstance.
● The orderly carrying out of the work was disrupted to a considerable extent.
● The loss is not covered by some other provision in the contract.

Two other contract clauses provide for the contractor to claim loss and/or expense:

● Clause 13.5.6 refers to a 'fair valuation' being made if a variation, in part or in whole, does not relate to additions, substitutions or omissions and the contractor cannot be reimbursed under any other provision of the contract. This appears to be a mopping up clause to ensure the valuation of Architect's Instructions which would not be covered by the remainder of clause 13. For example, a discrepancy between drawings and bills of quantities is to be rectified by an Architect's Instruction in accordance with clause 2.3. It is quite conceivable that the drawings might be amended while the work in the bills remained unaltered. Nonetheless, the contractor may suffer loss. Change to the order of the work (clause 13.1.2) is also covered.
● Clause 34.3.1 provides for the contractor to be paid if he suffers direct loss and/or expense following the discovery of antiquities and provided he cannot be reimbursed under any other provision of the contract.

1.5.2
ACA82

The principal clause allowing a claim for damage, loss and/or expense is clause 7. Clause 7.1 makes provision for the contractor to claim 'damage, loss and/or expense' incurred if the 'regular progress of the Works or of any section' is disrupted or the execution of such works are delayed in accordance with the dates stated in the time schedule due to any act, omission, default or negligence of the employer or of the architect.

The clause is extremely broadly drafted. The phrase *'damage, loss and/or expense'* might be expected to mean rather more than simply loss and/or expense, otherwise there would be little point in adding a superfluous word. However, there appear to be no grounds for giving the expression used in this contract any other meaning than that given to the expression *'loss and/or expense'* in the JCT80 form: in other words, damages as they are commonly understood. Damage, loss and/or expense is not qualified by the word *'direct'* as in the JCT80 form, but probably the words *'in consequence of'* have the same effect (see section 1.5.1). The phrase *'regular progress of the Works'* means exactly the same as in the JCT80 form (see section 1.5.1). However, the word *'disruption'* is a more forcible expression than the phrase *'materially affected'*, but in practice the meaning must be taken to be much the same—rather more than a small irritating disturbance.

The inclusion of the word *'delay'* simply makes it clear that the contractor may or may not be delayed in accordance with the time schedule. There may be disruption and delay, disruption without delay or simply (and less usually) delay alone. Each case qualifies for payment if caused by any act, omission, default or negligence of the employer or architect.

It is important to note that the broad drafting of the clause enables the contractor to apply to you for *all claims* arising out of any act etc of the employer or yourself, including what would normally be classed as ex-contractual claims, provided:

● They do not relate to Architect's Instructions.
● The recovery of the damage, loss and/or expense is in accordance with clause 7.2.

The clause, therefore, places a greater burden on you than clause 26 of the JCT80 form.

Two other clauses provide for the contractor to claim damage, loss and/or expense:

● Clause 10.3 provides for the contractor to be paid under the provisions of clause 7.2 if the regular progress of the works is disrupted or they are delayed in relation to the time schedule and the contractor suffers damage, loss and/or expense due to the execution of any work or installation of any materials etc by the employer or his employees, agents and contractors.
● Clause 17 refers to damage, loss and/or expense incurred by the contractor arising out of or in connection with certain Architect's Instructions. This, of course, is in addition to the valuation of any work included in an instruction. The phrase *'arising out of or in connection with'* appears to reinforce the contractor's right to consequential loss. On the other hand, he will have no claim if the instruction refers to matters which are to be *reasonably inferred* from the contract documents.

1.5.3
GC/Works/1
The principal clause allowing a claim for expense is clause 53. The contractor may claim if he 'properly and directly incurs any expense in performing the contract' because the 'regular progress' of the works is materially disrupted or prolonged due to a number of different circumstances.

'*Expense*' is expenditure incurred. It does not cover failure to obtain as much profit as was envisaged, except in the sense that increased expenditure would reduce profit. Loss of future profits is not covered. There are two provisos:

● The expense must not be such that the contractor would have incurred it irrespective of the circumstance occurring.

● The expense must not have been 'reasonably contemplated' by the contract. What might be reasonably contemplated will depend upon each contract, the drawings and bills of quantities.

The word *'properly'* means that the contractor must have been carrying out his duties in the correct way. The word *'directly'* means that consequential expense is not included. The expense must be the direct result of the circumstances without any intervening event (see section 1.5.1). The phrase *'regular progress of the Works'* means exactly the same as in the JCT80 form (see section 1.5.1), and the use of the words *'materially disrupted'* makes it clear that the result must be a substantial occurrence.

The inclusion of the word *'prolonged'* simply makes it clear that the contract may or may not be prolonged, but if either material disruption or prolongation or both are present a claim is possible.

Therefore, the contractor can claim if:

● One of the circumstances was the cause of the expense and not just an accompanying circumstance.
● The orderly carrying out of the works was disrupted to a considerable extent and/or the contract period is prolonged.
● The contractor was carrying out his contractual duties correctly.
● The expense is not covered by some other provision and it is not reasonably contemplated by the contract.

One other clause, i.e. clause 9(2)(a)(i), provides for the contractor to claim expense if he incurs expense in complying with Architect's Instructions (other than instructions for alterations, additions or omissions). In order to be able to claim under this clause, the contractor must satisfy the same four conditions noted above.

1.5.4
Agreement for Minor Building Works

There is no provision for the contractor to claim or for you to ascertain loss and/or expense (see section 10.5).

1.6 Circumstances that give rise to claims

A claim is not the same thing as a valid or a successful claim. A claim may be made by the contractor because:

● He is losing money.
● He wants more money.
● He is vexatious.
● The architect or employer is in breach or he can bring his claim within the 'claims' clause.

Only the last reason can be the starting-point for a valid claim. It must be remembered, however, that if the contractor is losing money, he may become very ingenious in formulating a claim. The ideal contract, where all parties work together amicably to produce the finished structure, helped by a set of competent drawings and bills of quantities, would not give rise to any claims. Valid claims can arise only because the contractor suffers loss through your fault or the fault of the employer or if the contract gives a specific claim because of some particular event. Fertile soil for the generation of such claims include:

● Discrepancies between drawings, schedules and bills of quantities.
● Late or hurried preparation of detailed information.
● Multiplicity of Architect's Instructions.
● Constant revisions to drawings.
● Changes of mind on the part of the employer.
● Poor co-ordination between two or more parties involved in the construction process.

It is probably true, although regrettable, that some of these items are present in most construction contracts. Although claims may not be made for a variety of reasons, you must be forever on your guard to do all things, which the contract requires of you, efficiently and at the correct time. If you fail, there is a breach of contract for which the employer will be responsible at common law. He will, naturally, look to you for redress.

**1.7
Contract clauses
that might support a
claim**

It has been said that two-thirds of the clauses of most contracts can form the basis of a claim. Of course, not all such claims would be contractual. Many would be simple claims for damages at common law. Tables 2, 3 and 4, which follow, are intended to indicate the clauses that might give rise to claims, the type of claim and who normally deals with it. A separate table is given for each of the contracts under consideration except the Agreement for Minor Building Works. The tables cannot be comprehensive because particular circumstances may create a feasible claim from the most unlikely clause.

Table 2   JCT80 clauses that may give rise to claims

Clause no.	Event	Type	Usually dealt with by
2.2.2.2	Departure from SMM or error in description or quantity	Contractual	Architect
2.3	Discrepancy between any two or more of: drawings; bills; and instructions (not being a variation)	Contractual	Architect
3	Failure to take ascertainment into account in next interim certificate	Ex-contractual	Employer
4.1.1	Additions, omissions, alterations of materials, workmanship or order of work	Contractual	Architect
4.1.2	Employer employs others to carry out work without going through procedure	Ex-contractual	Employer
4.3	Instructions by employer (or architect orally)	Ex-contractual	Employer
5.2	Failure to provide correct documents	Ex-contractual	Employer

Clause no.	Event	Type	Usually dealt with by
5.3	Documents purporting to impose obligations beyond contract documents	Ex-contractual	Employer
5.4	Failure to provide information as necessary	Contractual or ex-contractual	Architect Employer
5.8	Failure to send duplicate copies of certificates to contractor	Ex-contractual	Employer
6.1	Divergence between statutory requirements and contract documents in clause 2.3	Contractual	Architect
6.2	Statutory fees and charges	Contractual	Architect
7	Failure to provide levels or accurate dimensioned drawings for setting out	Ex-contractual	Employer
8.1	Materials, goods or workmanship not procurable	Contractual	Architect
8.3	Opening up of work or testing of materials or goods found to be in accordance with the contract	Contractual	Architect
8.4	Wrongly phrased instructions	Ex-contractual	Employer
8.5	Unreasonable or vexatious instructions	Ex-contractual	Employer
9.2	Royalties and patent rights	Contractual	Architect
12	Clerk of works exceeding duties	Ex-contractual	Employer
13.1.3	Nomination of a sub-contractor	Ex-contractual	Employer
13.2	Variations	Contractual	Architect
13.3	Expenditure of provisional sums	Contractual	Architect
13.4	Failure to value	Ex-contractual	Employer
13.5.1–13.5.5	Failure to value in accordance with the rules	Ex-contractual	Employer
13.5.6	Valuations not otherwise included	Contractual	Architect
13.6	Contractor not given the opportunity to be present at time of measurement	Ex-contractual	Employer
13.7	Failure to give effect to a valuation	Ex-contractual	Employer
15	Recovery of VAT or loss of input tax	Contractual	Architect

Clause no.	Event	Type	Usually dealt with by
16.1	Unreasonable withholding of consent to removal of materials and goods	Ex-contractual	Employer
17.1	Failure to issue certificate of practical completion	Ex-contractual	Employer
17.2	Failure to deliver a schedule of defects	Ex-contractual	Employer
17.3	Issue of instructions to make good defects after delivery of schedule of defects or after 14 days from the expiration of the defects liability period	Ex-contractual	Employer
17.4	Failure to issue certificate of making good defects	Ex-contractual	Employer
17.5	Instruction to make good damage by frost appearing after practical completion not caused by injury before practical completion	Contractual	Architect
18.1	Possession without contractor's consent	Ex-contractual	Employer
18.1.1	Failure to issue certificate of partial possession	Ex-contractual	Employer
18.1.2	Failure to give effect to all the results of the certificate	Ex-contractual	Employer
18.1.3	Failure to issue certificate of making good defects	Ex-contractual	Employer
18.1.5	Failure to reduce liquidated damages correctly	Ex-contractual	Employer
19.1	Assign by employer without consent	Ex-contractual	Employer
19.2	Unreasonably withholding consent to sub-letting	Ex-contractual	Employer
21.2	Special insurances under a provisional sum	Contractual	Architect
22A.4.2	Accepted insurance claims	Contractual	Architect
22B.2.2	Restoration of damaged work	Contractual	Architect
22C.2.3.3	Restoration of damaged work	Contractual	Architect
23.1	Failure to give possession of site on due date	Ex-contractual	Employer
23.2	Postponement of work	Contractual	Architect
24.2.2	Repayment of liquidated damages deducted.	Contractual	Architect

Clause no.	Event	Type	Usually dealt with by
	Interest	Ex-contractual	Employer
25	Financial claims	Ex-contractual	Employer
26	Loss and/or expense	Contractual	Architect
27.1	Invalid determination of the contractor's employment	Ex-contractual	Employer
28.2	Payment after contractor's determination	Contractual	Architect
29	Works by employer	Contractual	Architect
30	Certificates	Contractual	Architect
	Failure to observe rules	Ex-contractual	Employer
	Interest on retention	Ex-contractual	Employer
31	Failure to observe the requirements of this clause	Ex-contractual	Employer
32	Invalid determination of the contractor's employment	Ex-contractual	Employer
32.3	Payment for protective work	Contractual	Architect
33.1	Making good war damage	Contractual	Architect
34.3	Loss and/or expense in regard to antiquities	Contractual	Architect
35	Nominated sub-contractors	Ex-contractual	Employer
36	Nominated suppliers	Ex-contractual	Employer
38–40	Fluctuations	Contractual	Architect

Table 3 ACA82 clauses that may give rise to claims (*Note:* Due to the broad provisions of clause 7, all the events listed may be the subject of contractual claims. However, there is nothing to prevent the contractor pursuing any claim through the courts if he considers it offers him a better chance of success or more money. All the following claims would be dealt with by the architect)

Clause no.	Event
1.4	Mistake in contract bills
1.5	Discrepancy or ambiguity in drawings or documents comprising the contract documents
1.6	Compliance with statutory requirements
2.1 (alternatives 1 and 2)	Failure to supply drawings or details
2.3	Failure to deal with documents in due time
2.5	Incorrect drawings
3.2	Failure to comply with clause 2
6.4 (alternatives 1 and 2)	Accepted insurance claims
7.1	Damage, loss and/or expense
8.1	Instruction by employer
8.2	Architect's Instructions
9.1	Assignment by employer without consent
9.4, 9.6 and 9.7	Unreasonably withholding consent to sub-letting
10.3	Disruption of progress
11.1	Failure to give possession of the site on the due date
11.2	Failure to issue certificate
11.5 (alternative 1)	Failure to complete due to causes beyond the control of the contractor and liquidated damaged deducted
11.8	Acceleration or postponement
12.1	Failure to issue certificates in due time
12.3	Defective work due to employer or architect
12.4	Wrongful deduction
13.1	Taking-over without consent

Clause no.	Event
13.3	Failure to issue certificate within due time
16.2	Failure to issue certificate or failure to issue it in proper form
16.4	Failure to hold retention in separate account. Interest on retention
16.6	Failure to properly adjust the contract sum
16.7	Recovery of VAT
17.1	Variations Damage, loss and/or expense
17.3	Failure to act after time limit
18.1	Fluctuations
19.1	Failure to issue Final Certificate in proper form or at proper time
20.1, 20.3 and 21	Invalid termination
22.2	Payment after contractor's termination
23.1	Failure to properly serve notices

Table 4   GC/Works/1 clauses that may give rise to claims (*Note:* SO is the abbreviation used for the superintending officer, the architect for the purposes of this book.)

Clause no.	Event	Type	Usually dealt with by
5(2)	Error in bills of quantities	Contractual	SO
6	Failure to give possession of the site	Ex-contractual	Authority
7(1)(a)	Variation or modification of design, quality or quantity or addition, omission or substitution of any work	Contractual	SO
7(1)(b)	Discrepancy in or between specification, bill of quantities, drawings	Contractual	SO
7(1)(c)	Removal from site of goods for incorporation and substitution of other goods	Contractual	SO
7(1)(e)	Order of execution of any part of the work	Contractual	SO
7(1)(f)	Hours of working and extent of overtime or nightwork	Contractual	SO
7(1)(i)	Opening up for inspection of work found to be in accordance with the contract	Contractual	SO
7(1)(k)	Emergency work	Contractual	SO
7(1)(m)	Any instruction other than those in 7(1)(a)–7(1)(l) inclusive	Contractual	SO
9	Variations	Contractual	SO
9(1)	Failure to value in accordance with the rules	Ex-contractual	Authority
9(2)(a)(i)	Expense not otherwise included, in complying with instructions	Contractual	SO
12	Failure to supply dimensioned drawings, levels, or other information for setting out	Ex-contractual	Authority
13(3)	Tests on goods found to be in accordance with contract	Contractual	SO
15	Royalties and patent rights	Contractual	SO
21	Failure to examine and/or approve excavations for foundations	Ex-contractual	Authority
23	Making good defects due to frost or inclement weather not the fault of the contractor	Contractual	SO
26(2)(b)(i)	Damage to works or other things caused by the authority	Contractual	SO
26(2)(b)(ii)	Damage to works or other things caused by any of the accepted risks	Contractual	SO
28	Financial claims	Ex-contractual	Authority
28A(1)	Possession without the contractor's consent	Ex-contractual	Authority

Clause no.	Event	Type	Usually dealt with by
28A(3)	Failure to certify	Ex-contractual	Authority
28A(5)	Failure to release reserve	Ex-contractual	Authority
38	Prime cost sums	Contractual	SO
38(2)	Fixing goods	Contractual	SO
38(3)	Failure to adjust contract sum	Ex-contractual	Authority
39	Provisional sums	Contractual	SO
40(1)(2)	Failure to pay 97 per cent of value of work executed or goods brought onto site for incorporation	Ex-contractual	Authority
40(3)	Failure to value or certify in accordance with the rules	Ex-contractual	Authority
41(1)	Failure to pay the Final Sum less one half the reserve on Completion	Ex-contractual	Authority
41(3)	Failure to pay sum remaining due after certification of satisfaction and sum has been agreed	Ex-contractual	Authority
41(4)	Failure to pay in accordance with this clause	Ex-contractual	Authority
42(1)	Failure to certify	Ex-contractual	Authority
44(3)	Failure to operate provisions after determination	Ex-contractual	Authority
44(3) and 46(2)	Payment after determination	Contractual	SO
50	Works by authority	Ex-contractual	Authority
53	Prolongation and disruption	Contractual	SO

2 Roles

2.1
Architect

The contract places certain duties on you when you are dealing with contractors' claims. You are required to form an opinion and give effect to that opinion in a competent manner. In the course of your duties, you must take account of the facts and apply your knowledge and skill impartially.

For many years, the architect was believed to act in a quasi-arbitral capacity when carrying out his duties between employer and contractor. It was thought that he was immune from claims alleging negligence. Following cases before the courts (notably Sutcliffe *v.* Thackrah (1974)), it has become doubtful whether there is any occasion when the architect has immunity (unless specifically so agreed between the parties). The changing view of the architect's position does not remove your obligation to act impartially, but it does mean that you are liable for your actions. For practical purposes, you must be considered to be the agent of the employer, albeit in a limited capacity.

You are, therefore, in an invidious position. The person most likely to sue you for negligence is the employer. There are a number of situations when, acting impartially, you should find in favour of the contractor as a result of your own default; but you will do so in the almost certain knowledge that the employer will withhold fees or take legal action to recover his loss.

A simple example will illustrate the point. If you have to postpone part of the work under JCT80, clause 23.2, because of your inadequate preparation of drawings, the contractor may claim loss and/or expense under clause 26.2.5. Once the claim is ascertained and certified, the employer will demand an explanation for the additional cost. You will then be in the position of giving the employer evidence to use against you.

Of course, as a professional person, you are expected to exercise your skills properly and not make serious errors; if you do, you are expected to reveal your mistake and take the consequences. It is a burden you are required to shoulder. Having said that, it is clear that there will be some architects, faced with an adverse situation, who will do everything to reduce the contractor's claim and thus their own possible future liability. Put at its weakest, the tendency is for contractors to receive less than their due in such circumstances. This may be considered to be a weakness in the present system. It is not the purpose of this book to propose theoretical revisions, but to make the current situation clear and suggest ways of dealing with it.

You have two quite distinct functions in relation to claims:

● Extension of time: You must estimate a fair and reasonable extension of the contract period (with the approval of the authority in the case of GC/Works/1).
● Loss and/or expense: You must ascertain validity and ascertain (or instruct the quantity surveyor to ascertain) the amount payable.

It is important to recognise the difference between estimation and ascertainment. An estimate is an approximate judgement, not something to be finely calculated. An ascertainment is the finding out of something with great accuracy, and the end product must be certain.

In using different words to describe your functions, contracts recognise that the process of awarding an extension of time is subject to no precise formula and, in some cases, considerable gut feeling. Contracts make clear, however, that the calculation of loss and/or expense must leave no room for doubt. That is not to say, of course, that you may not be called upon, in either case, to substantiate your award before an arbitrator. How you can be in a position to do that will be dealt with in chapters 5 and 6.

2.2
Quantity surveyor

The quantity surveyor's job in regard to claims is very clear, thanks to recent case law (County and District Properties Ltd *v* John Laing Construction Ltd (1982)). He may calculate the amount of money due to the contractor if you so instruct, but it is not his function to decide liability. You receive the contractor's claim and may give your findings to the quantity surveyor who calculates the sum payable.

It is never appropriate for you to delegate the estimation of an extension of time to the quantity surveyor (although it is sometimes done) and there is no provision for doing so in the contract. It is, of course, perfectly reasonable for you to consult the quantity surveyor, if appropriate, before awarding an extension of time or deciding liability.

2.3
Consultants

Unlike the architect and quantity surveyor, the consultant has no specifically defined role in relation to contractor's claims. Indeed, there is no reference to consultants in the forms of contract under consideration. If the consultant is to be involved in claims at all, it will be at your invitation and only in the capacity of an expert witness, as it were, to lend an opinion from within his own area of expertise. For example, if the contractor wishes to claim an extension of time and loss and/or expense due to late receipt of heating information, you might, quite reasonably, ask the heating consultant if, in his opinion, the information was late and what would be the likely result in terms of delay or disruption.

However much you canvass opinion, the final decision is yours.

2.4
Clerk of works

Although the contract gives him no part to play in the determination of contractors' claims, he is very precisely noted in JCT80, clause 12, as being an inspector on behalf of the employer and under your direction. In this position he is uniquely placed to observe minute details of the job and to comment on the running of the job on site. His observations will be a major factor in the determination of any claim. If he is doing his work properly, you will be kept precisely informed throughout the contract of all you need to know about the progress of the contract.

The ACA form and the Agreement for Minor Building Works make no reference to a clerk of works. If one is appointed, his duties must be clearly stated and agreed between the parties. However, they are unlikely to extend beyond the clerk of works' duties under the JCT80 form and they will not include direct involvement in claims.

The GC/Works/1 form makes provision for a resident superintending officer (SO) or clerk of works, in clause 16. His duties are that of an inspector under clauses 2 and 13, which goes rather further than JCT80 clause 12. Although the contract provides for further powers to be delegated to him, it is not conceivable that they would include the consideration of claims.

2.5 Employer

It might appear cynical to say that the role of the employer is to pay, but in the case of contractual claims that is, in fact, the case. The employer has no power to interfere with your judgement and it would be most improper of him to try to do so.

In most contracts the employer is the only non-expert concerned, which can lead to problems. It is your job to explain the contractual provisions to him. Unfortunately, the employer sometimes finds it difficult to accept that, although you are his agent, you will not automatically try to defeat every claim made by the contractor. If the employer is a professional or a company used to commissioning the erection of buildings, the situation should not arise. In any case, you should give a full report to the employer as soon as you have made your decision.

The employer will have to take a more active role in the case of ex-contractual claims (see chapter 10).

3 Contractor's duties

3.1
Extension of time

The contractor's duties are shown in flow chart 1.
The contract lays down (clause 25) very precise
rules which the contractor must follow:

3.1.1
JCT80

● As soon as the contractor thinks he is being
delayed or is likely to be delayed in the future, he
must immediately notify you in writing. (Note that
the contractor is obliged to notify you of delays
even if there is no prospect of the contract period
being extended.) Verbal notification or
notification minuted in agreed site meeting
minutes is not sufficient. The notice must include
the reasons for the delay (even if it is the
contractor's own fault in part or in whole) and
state, in his opinion, which if any of the causes of
the delay is a *relevant event* (clause 25.4).
● As soon as possible, the contractor must give
details of the expected results of the relevant
events.
● As soon as possible, the contractor must
estimate the extent of the expected delay beyond
the contract completion date. He must note the
effect of each relevant event separately, stating
whether or not the delays will be concurrent. If he
estimates that there will be no overall delay to the
contract completion date, he must so inform you.
● If the contractor makes any reference to a
nominated subcontractor, he must send a copy of
the written notice, particulars and estimate of
delay to the nominated sub-contractor concerned.
● The contractor must give you further notices to
update the particulars of delay and estimate of the
effect on the completion date from time to time as
necessary or in order to comply with your
requests. He must send a copy of each such further
notice to any nominated sub-contractors who
received a copy of the first notice.

The contractor has a continuing obligation to notify delays throughout the contract period and until the works have reached practical completion. Failure to do so will put him in breach of the terms of the contract, quite apart from seriously affecting his chances of obtaining any extension of time.

It is in the contractor's own interests to present all the information he has. He must carefully document his notices to you and answer any requests for further information promptly. In effect, the contractor has to argue his case.

Besides providing written notices to you, the contractor has two further duties:

● He must constantly use his best endeavours to prevent delay in progress and prevent or reduce any effect upon the completion date. This does not mean that he should spend money to make up any delay but that he should be sure that he is proceeding diligently. Thus, if any part of the delay is his own fault, it can be interpreted as a lack of endeavour on his part, depending upon the circumstances.

● He must do everything that you reasonably require to proceed with the works. It is not reasonable for you to require the contractor to spend additional money without recompense to catch up on delays but, with this proviso, the contractor's clear duty is to follow your instructions with regard to progress.

These two duties are closely connected. If the employer requires and authorises acceleration measures together with appropriate payment, there is a duty upon the contractor to carry them out. However, under JCT80 terms, there is no power to order acceleration. The contractor's agreement is necessary.

3.1.2
ACA82   There are two sets of provisions for extensions of time—alternative 1 and alternative 2. The grounds for extension are different (alternative 2 is broader) but the contractor's duties are the same

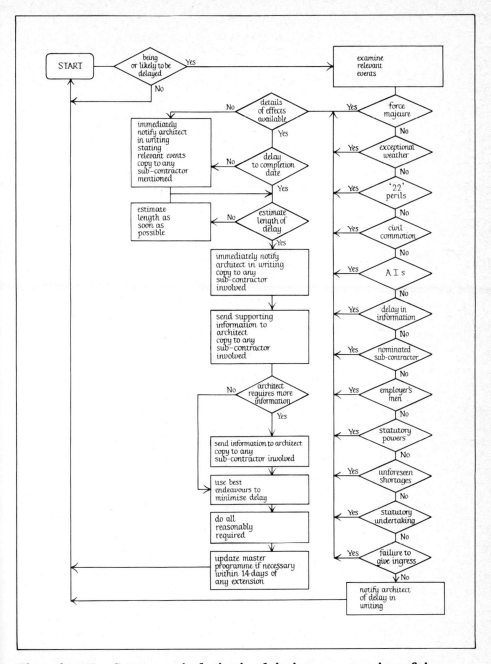

Flow chart 1   Contractor's duties in claiming an extension of time under JCT80

in each case. They are shown in flow chart 2. The contract lays down precise rules which the contractor must follow:

● As soon as the contractor thinks he is being delayed or is likely to be delayed in the future and, as a result, the taking-over of the works on the due date will be prevented by any of the events in clause 11.5 (alternative 1 or 2), he must immediately notify you in writing. It must be a specific notice; minutes of site meetings are not sufficient.

● The notice must specify the circumstances and give full and detailed particulars of the extension of time to which he considers he is entitled. The information must be as full and clear as possible.

● The contractor must give any further particulars which you may request to enable you to carry out your duties in estimating a fair and reasonable extension of time.

There is no obligation on the contractor to notify delays in general. He need only notify them if he wishes to claim an extension of time.

The contractor's duties are qualified in respect of *any act, instruction, default or omission* of either the employer or yourself. He is still contractually bound to notify you as stated above but should he fail to do so, you still have the power to act. This is because failure to give an extension of time for defaults etc of the employer or his agents would render the liquidated damages clause (clause 11.3 alternative 1) unenforceable. It would almost certainly have the same effect upon the unliquidated damages clause (clause 11.3 alternative 2). The contractor should take heed, however, that failure to give notice allows you to delay granting an extension until immediately before issuing a final certificate. It is, therefore, very much in the contractor's own interests to carry out all his duties precisely and at the proper time.

The contractor has three further duties:

● If you issue an instruction under clause 11.8 to bring forward or postpone the dates for taking-over of any part of the works (there is no provision for the taking-over of the whole of the works to be brought forward or postponed), the contractor must immediately take measures to comply.

● Within fourteen days of the date of your notice of extension under clause 11.6 or your instruction under clause 11.8, the contractor must submit to you a revised time schedule for your consent.

● There is no specific requirement for the contractor to use his best endeavours, but clause 11.1 requires him to proceed regularly and diligently and in accordance with the time schedule.

3.1.3
GC/Works/1
The contractor's duties are shown in flow chart 3. The contract lays down precise rules which the contractor must follow:

● As soon as the contractor thinks the completion date is being delayed or is likely to be delayed in the future by any of the circumstances in clause 28(2), he must immediately notify the architect in writing. It must be a specific notice, and minutes of site meetings are not sufficient.

● The notice must specify the circumstances and the extent of the delay caused or likely to be caused. The information must be full and clear.

There is no express obligation for the contractor to provide further particulars at your request, but it would be a foolish contractor who did not do so because he would risk having his claim rejected for lack of evidence.

There is no obligation upon the contractor to notify delays in general. He need only notify them if he wishes to claim an extension of time.

Note that the right of the contractor to an extension of time depends upon his strict observance of the above duties *unless the authority decides otherwise*. The authority is likely to waive the contractor's duty only in the case of an act or

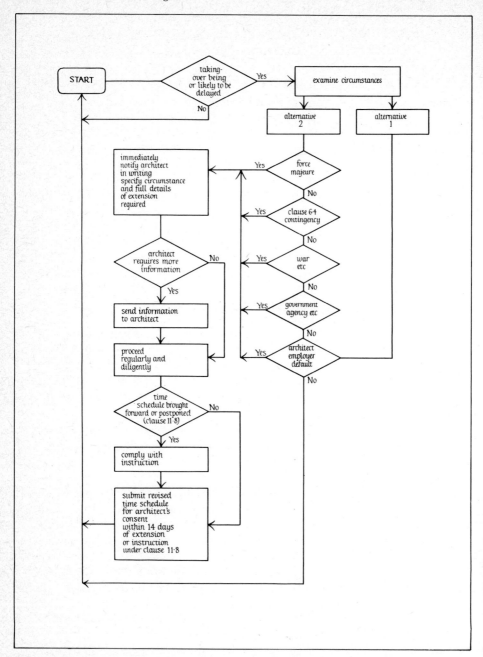

Flow chart 2   Contractor's duties in claiming an extension of time under ACA82

default of the authority, in order to prevent the date for completion becoming at large.

The contractor has two further duties:

● He must always use his best endeavours to prevent or minimise any delay caused by any of the circumstances in clause 28(2). This does not mean that he should spend money to make up any delay, but that he should be sure that he is proceeding diligently. Thus, if any part of the delay is his own fault, it can be interpreted as a lack of endeavour on his part, depending upon the circumstances.

● He must do all that may reasonably be required, to your satisfaction, to proceed with the works. It is not reasonable for you to require the contractor to spend money without recompense to catch up on delays but, with that proviso, the contractor's clear duty is to follow your instructions with regard to progress.

The two duties are closely connected. If the authority requires and authorises acceleration measures together with appropriate payment, there is an obligation upon the contractor to carry them out.

3.1.4
Agreement for Minor Building Works

The contractor's duties are very simply stated in clause 2.2:

● As soon as the contractor realises that the works will not be completed by the due date, he must inform you.

Although he has no further obligation, he should be ready to furnish additional details to help you to decide upon a reasonable extension of time.

3.2
Loss and/or expense

3.2.1
JCT80

The contractor's duties are shown in flow chart 4. Clause 26 sets out the contractor's duties if he wishes to claim loss and/or expense. Unlike the duties under clauses 25 and 35.14 and clause 11 of the sub-contract NSC/4 or NSC/4a, which are

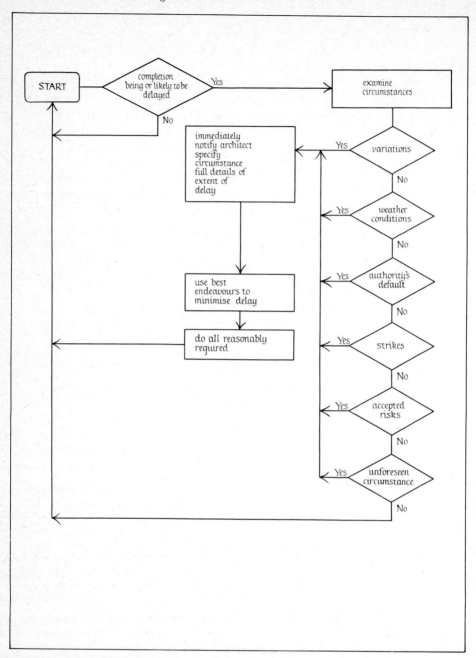

Flow chart 3   Contractor's duties in claiming an extension of time under GC/Works/1

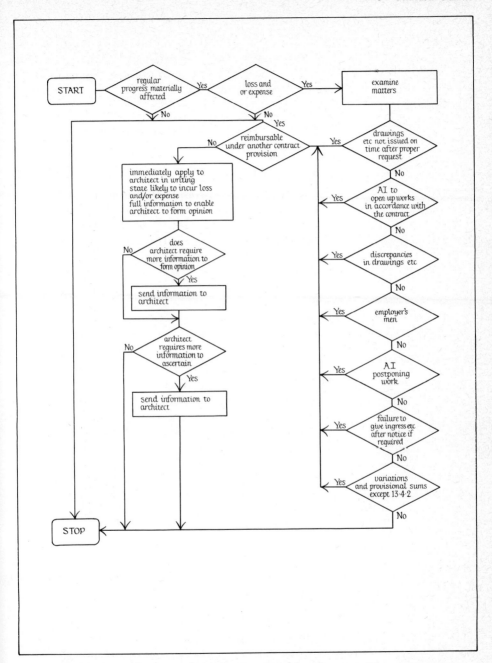

Flow chart 4   Contractor's duties in claiming loss and/or expense under JCT80

mandatory whether or not the contractor is entitled to an extension of time, most of the duties under clause 26 are operative only if the contractor wishes to claim loss and/or expense. If he does wish to claim he must:

● Apply to you in writing.
● Apply as soon as it becomes apparent that the regular progress has been or is likely to become affected.
● State that he has or is likely to incur loss and/or expense in carrying out the contract, for which he would not be reimbursed under any other provision of the contract; because the works have been or are likely to be substantially affected by one or more of the matters in clause 26.2.
● Supply, on request, information to enable you to form an opinion.
● Supply, on request, whatever details are reasonably necessary to enable ascertainment to take place.

The type and extent of information necessary to enable you to form an opinion and then to carry out ascertainment is often the subject of bitter argument. It is usual, and wise, for the contractor to submit a full and detailed report in support of his claim. It is against his own interests to refuse to supply any further details that might be required. Of course, he may reach the stage where he feels that he is being asked to supply additional information purely to delay the time when a decision or ascertainment must be made. However, he must be very sure that he has done all that is required of him before he refuses to submit further details or he may seriously prejudice his chances of obtaining payment.

Clause 26 is valuable to the contractor because it is specifically stated to be in addition to any other rights or remedies he may possess. In other words, the contractor may make a claim for damages under common law if he feels that it will provide him with a greater sum than he would otherwise get, or if he has failed to comply with the procedural provisions of clause 26. The contractor's duties under clause 26 have, of course, no direct bearing on any claim he makes through the courts where each claim is judged on its merits. He may, for example, pursue such a claim long after his contractual right has elapsed.

The contractor has no specific duties under clause 13.5.6 other than to provide the required information to enable a decision to be made.

The contractor has no duties under clause 34.3.1 but, in his own interests, he should co-operate in providing any information you require.

3.2.2
ACA82
The contractor's duties are shown in flow chart 5. Clauses 7 and 17 set out his duties in detail. If he wishes to claim under the provisions of clause 7 he must:

● Give you written notice as soon as it becomes apparent that an event has occurred or is likely to occur which will give rise to a claim.
● Include a description of the event in the notice.
● Within twenty-eight days of the notice, submit a detailed estimate of any adjustment to the contract sum to which he thinks he is due as a result of the event. All relevant supporting evidence must be included.

Presumably, on receipt of the contractor's notice, you will inform him what information is required so that he can supply it within the twenty-eight days deadline. There is a potential difficulty because, until you and the quantity surveyor see his estimates, you may not be able to say precisely what you require. It is envisaged that considerable activity will take place during the twenty-eight days. Initially, you should couch your letter in broad terms (see letter 2). If the contractor fails to carry out his duties, your power to adjust the contract sum and the contractor's rights to payment in respect of the claim are removed until the issue of the final certificate.

If the contractor wishes to claim in accordance with clause 17 (Architect's Instructions), he must give you, within fourteen days of receipt of the instruction:

● Estimates of the value of the instruction, together with all necessary supporting calculations.
● Estimates of any extension of time he thinks may be due under clause 11.5.
● Estimates of any damage, loss and/or expense incurred due to the instruction.

The contractor must not comply with the instruction meanwhile.

The time allowed is short, the information required could be extensive. You have power (clause 17.5) to dispense with the contractor's obligations to provide all the above information but you must then ascertain and certify a fair and reasonable adjustment to the contract sum. It is hard to envisage that the contractor will find such a procedure satisfactory.

If you do not dispense with his obligations and he fails to comply with them, your power to adjust the contract sum and the contractor's right to payment in respect of the claim are removed until the issue of the final certificate.

3.2.3
GC/Works/1

The contractor's duties are shown in flow chart 6. Clauses 53 and 9(2)(a) set out his duties in detail. If he wishes to claim under the provisions of clause 53, he must:

● Give you written notice as soon as he becomes aware that regular progress has been, or is likely to be, disrupted or prolonged by one or more circumstances in clause 53(1).
● Specify the circumstances in the notice and state that he is, or expects to be, entitled to an increase in the contract sum thereby.
● As soon as possible, provide all the documents

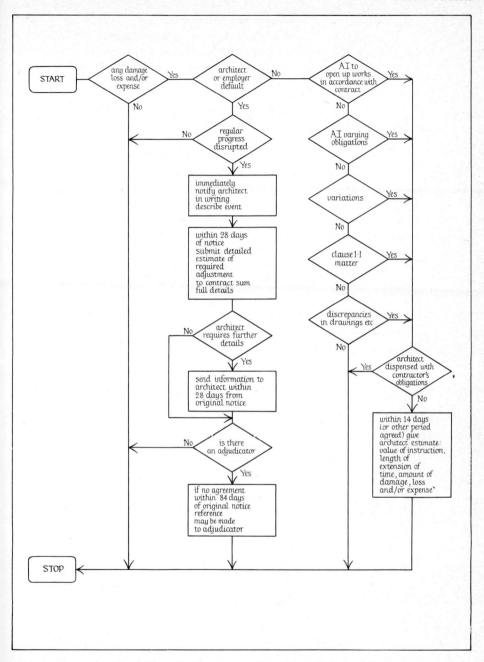

Flow chart 5   Contractor's duties in claiming damage, loss and/or expense under ACA82

2   Letter from architect to contractor requiring further information
This letter is suitable only for use with ACA82

Dear Sir,

Thank you for your notice of the [*insert date*].
I should be pleased to receive your estimate of
the adjustment to the contract sum which you
require; to take account of damage, loss and/or
expense, supported by documents, vouchers and
receipts necessary for computing the adjustment.

It is in your interests to send me the fullest
possible information in good time to allow me to
examine it and, if necessary, request and receive
such further information as I may require before
expiry of the 28[1] days deadline imposed by the
contract.

Yours faithfully,

Copy: Quantity Surveyor.

1. *Or substitute such other period as is inserted
   in the contract.*

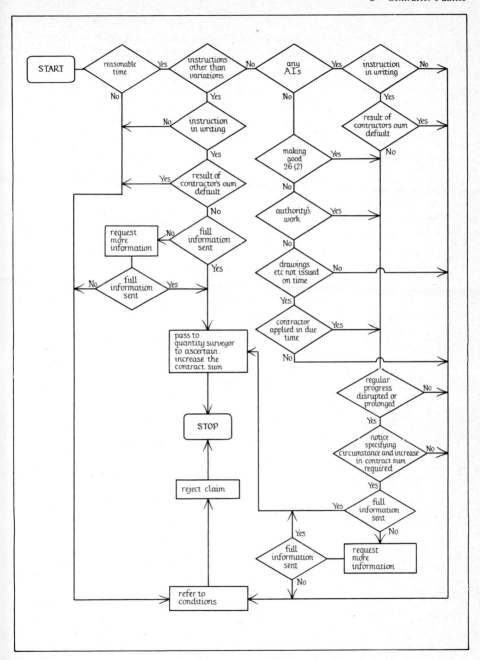

Flow chart 6   Contractor's duties in claiming expense under GC/Works/1

and information necessary to calculate the amount of expense as required by the quantity surveyor.
● Prove that any instruction was given or confirmed in writing.
● Prove that the items in clause 53(2), drawings, authority's direct works, nominations etc, were requested at a reasonable time in relation to the date they were required, unless a specific date was agreed.

The contract states that the contractor's duties are a condition precedent. In other words, if the contractor does not carry them out precisely, he has no right to payment and you and the quantity surveyor have no power to consider the claim.

If the contractor wishes to claim under the provisions of clause 9(2)(a), he must:

● Prove that any instruction was given or confirmed in writing.
● As soon as possible after incurring the expense, provide all documents and information necessary to calculate the amount of expense as required by the quantity surveyor.
● Prove that the instruction was not given as a result of his default.

The contractor's duties are again expressed in the contract as a condition precedent with all that implies (see above). Note that the contractor is not required to give any specific notice under this clause.

4 Evidence

4.1
Introduction

In submitting a claim for either extension of time or loss and/or expense, the contractor will normally produce evidence in support. You may require additional evidence on certain points before you can properly exercise your duties under the contract.

There are a number of misconceptions, on both sides, about what is, and what is not, good evidence. If the contractor simply presents you with a statement of the profit he expected and the loss he sustained with a claim for the difference (it has been done) it is not evidence at all. It is simply a vague and unsupported claim. The purpose of this chapter is to examine the various types of evidence which may be produced (see table 5) and to explain what value you should give it when forming a judgement.

4.2
The contract

This is the most important evidence regarding what was agreed between the parties. In some instances, however, you may need the advice of a lawyer to decide what it means.

4.3
Bills of quantities

When comparing bills of quantities with the drawings, it is possible to pin-point discrepancies. When priced, they give a clear picture, not only of the amount of work for which the contractor tendered but also of his pricing strategy. They may reveal, for example, that the contractor priced in a certain way in shrewd anticipation of a claim. Such a situation does not invalidate the claim, but it does provide a useful indication of the contractor's attitude. He could be truly said to be 'claims conscious'.

4.4
Specification

This is another primary piece of evidence that is common to both parties and, therefore, it cannot be in dispute, although the meanings of its provisions can.

4.5
Drawings

A copy should be kept of all drawings and schedules issued, for future reference. The evidence of a drawing is often crucial. If revisions have taken place, the date and nature of the revision should be noted on the drawing. The drawings provide clear evidence of your intentions and the quality of your instructions to the contractor. It may often be difficult for you to form an unbiased judgement on the standard of your own drawings. Ask the following questions:

● Is the information on the drawing capable of more than one interpretation? In particular, can it reasonably be interpreted as the contractor asserts?
● Is the subject of the complaint spread over two or more drawings? If so, is there any conflict between the different drawings (no matter how small)?
● Are the drawings unclear and are the notes difficult to read? (Test them on your colleagues, you may have an eccentric lettering style.)
● Were some of the drawings, relating to a disputed item, issued at different times?
● Do any drawings, which might reasonably be expected to show the information, not do so?

If the answer to any of the questions is 'Yes', concede—to yourself at least—that the contractor may have a valid point.

4.6
Correspondence

Letters are an extremely valuable source of evidence because they indicate intentions and attitudes at the time of writing. Letters must be read with care, however, and always in the context of other letters and documents. They may have been written especially to support a future claim. They may not mean what they appear to mean at first sight. Written evidence is always useful provided that it is clear. A letter from the contractor asserting that a particular situation exists (for example, lack of drawings) is persuasive in the absence of a letter from you refuting the assertion. Lack of refutation cannot be taken to mean agreement, although it may be suggestive, but a simple acknowledgement can.

Correspondence, written by one person for the eyes of another, is potentially the best sort of evidence save only for documents (such as the contract) agreed and signed by both parties.

4.7
Minutes of site meetings
The evidence of minutes is excellent provided that it is minuted as agreed in the minutes of a subsequent meeting. Care must be taken to check all correspondence and other notes following the date of the circulation of minutes to ensure that there is no dissent registered but omitted from the minutes. Minutes that are not agreed have a reduced value, somewhat like your own notes.

4.8
General reports
Reports are of two basic kinds:

● Information provided from one person or organisation to another (for example, your report to a client).
● Information provided within, and for the use of, a particular organisation (for example, a report called for by the senior partner of a practice from the project architect).

The value of a report as evidence will depend upon:

● The quality of information it contains.
● The person or organisation preparing it.
● The person for whom it is intended.
● The circumstances generating the report.

For example, some reports are prepared in order to explain away a mistake. The subsequent value of such a report may well be limited. A report for circulation within an organisation is more likely to put the facts bluntly than one intended for outside consumption. Ideally, all reports should be straightforward and clear. A contractor's internal report on an occurrence should not vary from his report to you on the same occurrence. In practice, individuals try to put the best possible interpretation on their actions, especially if money or reputation is at stake. There is often no written response to a report and great care must be used in evaluating its worth.

**4.9
Contractor's master programme**

The value of the contractor's programme depends upon:

● The type of programme.
● Whether it is a realistic forecast of the progress of the job.
● The accuracy of the monitoring process.

Some programmes are much better than others for revealing delays and disruptions:

● Network analysis and precedence diagrams are excellent. They give a very detailed picture of the intention, when compared to the reality that can be obtained.
● Bar charts are limited in scope and do not permit fine tuning.
● Lines of balance are theoretically good, but rather confusing in practice.

Many contractors produce unduly optimistic programmes almost as a matter of course, like a man setting his watch 10 minutes fast. The danger is similar. The man is always 5 minutes late and the contractor overruns. The security of a known buffer may induce a false sense of security. This type of programme is virtually useless as evidence of anything except, perhaps, the contractor's doubtful efficiency. A carefully worked out programme, on the other hand, methodically updated and monitored can be a key factor in deciding a claim.

**4.10
Clerk of works'
reports**

If a clerk of works is employed upon a job, he will normally produce a report for you each week. It will contain:

- Details of the visitors to the site.
- The number of operatives working each day in each trade.
- The state of the weather each day.
- The progress of the work.
- A brief written report on points of importance.

Frequently, other information is included to suit the requirements of a particular architectural practice (for example details of the drawings and architect's instructions received). It is a particularly useful account of the job in all its aspects and often forms the most useful evidence you will consider.

**4.11
Clerk of works'
diary**

All clerks of works should keep a daily diary, which provides greater detail of site activities than the report sheet, including defects and problems generally. Read together with the clerk's of works reports, the diary should form an accurate account of the situation on site throughout the contract period.

**4.12
Diary of the
person-in-charge**

This is a most valuable piece of evidence if the contractor can be persuaded to release it. There is no obligation upon him to show it to you. He is only bound to satisfy you of the validity of his claim. He may refuse to release the site diary for reasons totally unconnected with the claim (for example, the person-in-charge may have penned some extremely rude remarks about you or the clerk of works).

Very often, the contractor will provide photocopies of pages from the site diary. They must be treated with caution because:

- They may not be the original pages.
- They may be the original pages with parts blanked out or added.

51

● They may have an entirely different meaning if read in the context of the diary as a whole.

The moral is simple. If you are offered photocopies of the site diary, you should insist on seeing the original and, at least, check that the copy is accurate. Alterations are not easy to disguise in the original diary.

4.13
Architect's notes

You will probably have a great many notes in your file, including notes of site visits and telephone conversations. Although not conclusive in the way that a court would understand it, they may well be conclusive to you and serve to remind you of your actions or thoughts at particular stages in the contract. Notes will include your diary notes and notes put on letters. This last habit, incidentally, is extremely ill-advised for the simple reason that if the letter is produced in evidence at an arbitration hearing or in court the note is revealed with it. It may be embarrassingly rude or reveal something you wish to conceal. 'No chance!', scrawled across the top of a contractor's letter may be an accurate reflection of your mood at the time, but it will not convince a court of your impartiality. It is common for a partner in a practice to wish to communicate some thoughts about a letter to a project architect, but it is best done on a separate slip of paper pinned to the letter. The architect can at least claim some privilege over such notes if the need arises in a 'discovery' situation in connection with litigation or arbitration.

4.14
Documents from the contractor's office

The contractor may produce many different kinds of evidence from his own office including:

● letters to and from sub-contractors and suppliers
● letters to and from others
● record drawings
● site photographs
● orders
● invoices

- delivery notes
- build-up of costs.

Unlike a court of law or an arbitration hearing, there is no legal requirement for the contractor to give complete discovery of all relevant documents. Therefore, he may well reveal only those papers which support his case, suppressing those which do not. You have no real way of knowing what other information exists. You can ask to see certain papers, but you may be told that the business was done by telephone or that the papers are mislaid. There is no method of discovering the truth except by calling the contractor's bluff—which may be expensive and is seldom justified.

The value depends upon the completeness of the documents. Convenient gaps may be genuine, but they must be always viewed with suspicion.

4.15
Documents from
sub-contractors and
suppliers

Evidence from sub-contractors or suppliers should be obtained through the main contractor. If he refuses to supply the information, he could be in breach. Depending upon the circumstances, it could be reasonable for you to seek the information direct. Its value depends upon its completeness and the way in which it was obtained. Was it volunteered, given in answer to query, or supplied unwillingly?

4.16
The site

Theoretically, the site itself is primary evidence. The value of the evidence on site will depend upon the nature of the claim and how quickly you can see the site after the event. In many instances, the site itself will not provide any worthwhile evidence to assist you in assessing a claim.

4.17
Oral evidence

Oral evidence must always be treated with extreme caution. It is useful for confirming points that may be in the balance but it is unwise for you to rely on oral evidence alone. The problem is that memories are not reliable and all those in a

Table 5   Value of evidence in dealing with a claim

Note: Value is given on a scale of 1–3, 3 being the highest value

Evidence	Value	Remarks
Contract	3	Primary evidence
Bills of quantities	3	Primary evidence
Specification	3	Primary evidence
Drawings	3	Primary evidence
Correspondence	1–3	Depends on context
Site minutes	1–3	Depends if agreed and who prepared
General reports	1–2	Depends for whom prepared
Master programme	0–3	Depends on type and if realistic
Clerk's of works reports	1–3	Should be high value
Clerk's of works diary	2–3	Should be high value
Diary of person-in-charge	0–3	Depends on access to original
Architect's notes	3	Primary evidence to architect
Documents from contractor	0–2	Depends on completeness
Documents from sub-contractor	0–2	Depends on completeness
Site	0–3	Depends on circumstances of claim
Oral evidence	0–2	Depends on source

position to make comments about the contract can be grouped roughly either for or against the contractor.

Oral evidence usually relates to what occurred on site or what one person said to another. This type of evidence is not usually submitted by the contractor. It is more likely that you will gather it in an attempt to confirm various doubtful points.

Likely people to give you oral evidence are:

● your office colleagues
● consultants
● the quantity surveyor
● the clerk of works
● nominated sub-contractors or suppliers.

You must bear a number of points in mind:

● the credibility of the witness

● the reliability of the witness
● the effect of the evidence, if accurate, on the claim
● whether the evidence strengthens or weakens any other evidence.

Unless the circumstances are truly exceptional, it is always wise to disregard oral evidence that conflicts with other evidence. It should be the last source of your information.

5 Techniques for dealing with extensions of time

5.1
Introduction

5.1.1
JCT80

Clause 25 sets out precisely how you are to deal with contractor's claims for extension of time. The contractor's duties have been considered in section 3.1.1 and flow chart 7 sets out your duties.

You have 12 weeks in which to make a decision, from receipt of the last of 'reasonably sufficient particulars and estimate' from the contractor. If there is less than 12 weeks to run to the completion date, the decision must be made before completion date. This last requirement may be difficult to achieve if the notice, information etc lands on your desk only a week or so before completion date. Although the requirement is qualified by the phrase 'if reasonably practicable', you must do your best to comply and it is worth considering issuing the extension in two parts:

● The first extension, based upon the initial examination of the information (if an extension appears to be warranted).
● The second extension after a more detailed examination.

In such a case, you should let the contractor know what you are doing when you issue the first extension (see letter 3). If you are at all doubtful whether any extension is justified, you must carefully examine all the evidence first, as described later in this chapter, even if it means that you are unable to issue any extension until after the completion date. Provided that you act as quickly as possible and let the contractor know what you are doing (see letter 4) there should be no contractual problems. It is unreasonable to expect you to form an opinion in a week or even two (unless the matter is very simple) when the

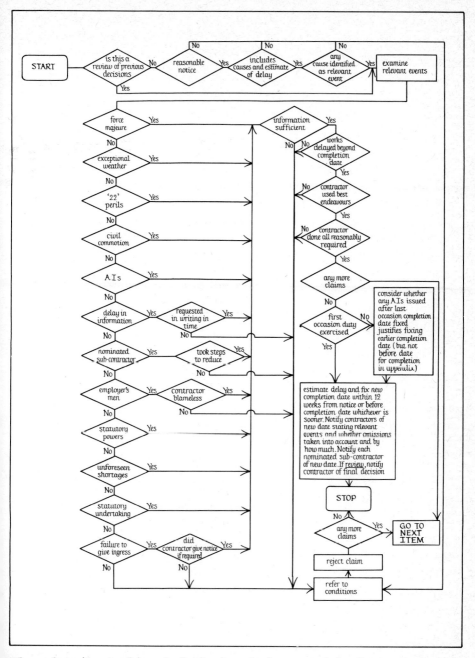

Flow chart 7   Architect's duties in relation to a claim for extension of time under JCT80

contract clearly envisages that you may take up to 12 weeks over the process. Make sure that you award the extension before practical completion, however, or you will get tangled up in the review you have to make.

After practical completion, you have a further 12 weeks during which you must review the extensions of time already given. You must then either:

- fix a later completion date *or*
- fix an earlier completion date *or*
- confirm the existing completion date.

There is an important difference between the consideration of extension of time before and after practical completion. Before practical completion, you must consider only the representation made, in proper form, by the contractor. Afterwards, you are not fettered and you are required to take account of any 'relevant events' and any instructions omitting work whether the contractor has specifically notified you or not.

If the works are uncompleted by completion date, you are obliged to issue a certificate to that effect (clause 24) and the employer may proceed to deduct liquidated damages. In the employer's interest, therefore, you should notify him (see letter 5) that further extensions may be due when you finish your review, otherwise the employer may have to refund the liquidated damages (possibly, although it is by no means firmly established, plus interest).

On receipt of a written notice from the contractor that the works are likely to be delayed, you are not obliged to do anything at all, even if the contractor has identified relevant events. Not until the contractor gives particulars of the effects of any relevant event and an estimate of the expected delay to completion of the works does your duty to estimate the amount of any extension of time begin.

The first thing you must do is to decide whether

3   Letter from architect to contractor if issuing extension in two parts
This letter is suitable only for use with JCT80 and GC/Works/1

Dear Sir,

I have received your submissions on the [*insert date*] in support of your claim for extension of time. In view of the proximity of the contractual completion date and the length of time it will take to consider all aspects of your claim, I enclose an interim award based upon my initial view of the evidence.

Any further extension of time to which you may be entitled will be awarded as soon as my investigations are concluded.

Yours faithfully,

3

4 Letter from architect to contractor if unable to issue extension before completion

Dear Sir,

I received your submissions on the [*insert date*] in support of your claim for extension of time. They are being considered as a matter of urgency.

In view of its proximity, it appears doubtful that I will be able to arrive at a decision until after the date for completion of the works.

Yours faithfully,

5   Letter from architect to employer if work not completed by due date
but further extensions may be due
This letter is suitable only for use with JCT80

Dear Sir,

I enclose my certificate in accordance with clause
24.1 of the contract.

You may, if you wish, deduct liquidated damages at
the rate stated in the Appendix for the period
between the date the contract should have been
completed and practical completion date.
Alternatively, you may recover the damages as a
debt.

Please bear in mind, when deciding whether or not
to deduct liquidated damages, that I have yet to
carry out my review of extensions of time. I
cannot begin until practical completion has
occurred. Further extensions may be due. If, as
a consequence of my review, I fix a later date for
completion, you would be liable to repay any
liquidated damages (and possibly interest also)
deducted for the period up to the later
completion date.

Yours faithfully,

5

the events noted by the contractor are indeed relevant events. In order to do this, you will usually require further information. In carrying out your duty, it is highly likely that you will have to contact the contractor on a number of occasions; each time seeking more information or clarification. The contractor may become impatient or even annoyed but, provided your requests are reasonable, he must furnish the data required. Only when you have the last of the particulars you need will the 12 weeks time limit begin to run.

Relevant events   It is timely to consider the relevant events in clause 25 and decide what they mean in practical terms. You must have a very clear understanding of the events which may be considered 'relevant' and those which may not. The relevant events are of three types:

● The fault of the architect or employer.
● The fault of third parties.
● Outside the control of the employer or contractor.

CLAUSE 25.4.1: Force majeure

Of all the relevant events, the first seems to cause the most trouble, simply because few people understand what it means. The usual interpretation is 'Act of God', but that expression is almost equally obscure. Many of the situations which normally would be covered by force majeure, such as strikes or the outbreak of war, are already provided for in the contract. Therefore, force majeure is best considered as a clause to catch events *not otherwise covered* which are outside the control of the parties to the contract and which they did not contemplate when they entered into the contract. In practice, force majeure may be much quoted by the contractor but events seldom stand up to investigation.

CLAUSE 25.4.2: Exceptionally adverse weather conditions

Contractors commonly serve notice and claim an extension under this clause at the first fall of snow or the first week of rain. Such a claim would be quite unacceptable. The clause refers only to *exceptionally* adverse weather. As well as severe frost or snow, many weeks of drought could qualify. The criteria are:

● Whether the conditions are extremely unusual for the time of year or location.
● Whether the contractor could have reasonably foreseen the conditions when he tendered.

Both criteria must be satisfied and the contractor would have to produce evidence, perhaps by means of meteorological reports, to support his contention. Snow in January is not unusual, but 10 weeks of continuous snow through December, January and into February may well be. Twenty years meteorological reports would be very good evidence as to the unusualness of the conditions and the degree of foreseeability involved.

CLAUSE 25.4.3: Loss or damage by clause 22 perils

This clause is straightforward and should pose little difficulty. Clause 22 perils are listed in clause 1 of the contract.

CLAUSE 25.4.4: Civil commotion, strikes etc affecting trades employed on the works or engaged in preparation, manufacture or transportation of goods etc

Take care in interpreting this clause. A civil commotion must be a substantial riot and while a strike (presumably official or unofficial) qualifies, a 'work to rule' does not. Moreover, these events must directly affect the trades. Thus a strike that affects a trade that in turn affects another trade involved in the works, does not qualify.

CLAUSE 25.4.5: Compliance with Architect's Instructions regarding discrepancies, variations, postponement, antiquities, nominated suppliers and the opening up of work found to be in accordance with the contract

It should be clear whether the event falls into any of these categories.

CLAUSE 25.4.6: The contractor not having received at the proper time necessary instructions etc for which he specifically applied at a reasonable time in relation to the date on which they were required

This clause may appear straightforward, but in reality it causes problems. Unless you know the contractor's detailed ordering and programming arrangements, you will find it very difficult to arrive at a fair judgement. A programme based on network analysis is invaluable in making a decision. You will need to have evidence of such things as the supplier's delivery period and the contractor's labour commitments. You must also take account of your own work programme. If the contractor forgets that he needs a particular drawing until two days before he is to construct the detail on site, there is no way that he can give you reasonable written notice. The important thing is that you cannot arbitrarily decide, for instance, that 4 weeks is sufficient time between request and requiring information, without considering every aspect of a particular claim.

CLAUSE 25.4.7: Delay on the part of nominated sub-contractors or suppliers which the contractor has taken all practicable steps to avoid or reduce

This clause has been roundly criticised in the JCT63 form (Jarvis and Sons Ltd *v* Westminster City Corporation (1970)) but it is reproduced word for word in the JCT80 form. Delay 'on the part of' is not the same thing as delay *caused by*. If the nominated sub-contractor's works contain defects that have to be corrected later or if he totally abandons the work for any reason, the contractor has no claim for extension of time. The only ground for claim is if the nominated sub-contractor does not complete by the sub-contract completion date (or the nominated supplier fails to deliver in accordance with the terms of his contract of sale). You must also consider carefully whether the contractor has taken all practicable steps to avoid or reduce the delay. If he has not, he is not entitled to any extension. A half-hearted letter of reproof from contractor to nominated sub-contractor does not qualify as 'all practicable steps'.

CLAUSE 25.4.8: The execution or failure by the employer to execute work not forming part of the contract in accordance with clause 29 or the supply or failure by the employer to supply materials or goods which he has agreed to supply

This clause is clear enough. Work or materials which the contract leaves for the employer to arrange will give the contractor grounds for extension if they cause delay. Included in this work is work carried out by statutory authorities *not* in pursuance of their statutory obligations and paid directly by the employer (Henry Boot Construction Ltd *v* Central Lancashire New Town Development Corporation (1980)).

CLAUSE 25.4.9: The exercise of any statutory power by the British Government, after the date of tender, which directly affects the works by restricting labour or materials essential to the works

But for this clause, such events would probably fall under force majeure. The occurrence should be obvious and *direct* (i.e. not the cause of a cause).

CLAUSE 25.4.10: The contractor's inability, for reasons outside his control and not reasonably forseeable at the date of tender, to secure labour or materials essential to carry out the works

Note that the contractor's inability to secure labour or materials is not, by itself, enough. Perhaps he did not make proper enquiries before he tendered. Not only must the problem be unavoidable, the contractor must be able to show that he could not have known, with reasonable enquiry, that it would be unavoidable. The contractor cannot claim twice for such events (i.e. under this clause and the preceding one).

CLAUSE 25.4.11: The carrying out of work, or failure, in pursuance of its statutory obligations, by a local authority or statutory undertaker in relation to the works

There is no qualification in this clause regarding the contractor's prior knowledge but clearly, if he knows a delay is likely he must take all reasonable steps to avoid it (clauses 25.4.1 and 25.4.2). This clause relates to the laying of pipes and cables and any other work which the authority is obliged to carry out in relation to the contract works—no other. The contractor is expected to include such work in his programme and, if a delay occurs, he can claim an extension.

CLAUSE 25.4.12: Failure of the employer to give, at the appropriate time, ingress to or egress from the site or any part; including necessary passage over adjoining land in the possession of the employer after the contractor has given any notice required by the contract documents or failure of the employer to give ingress or egress as agreed between the architect and the contractor

In order to be able to claim, the contractor must show:

● He has given any notice required.
● It is the employer's, or his agent's, fault.
● The employer is in possession of and controls the land in question.

Thus, for example, if road works block access, the

contractor has no claim under the contract because it is neither the fault of the employer nor does he control the road. The employer's failure to give access is not the same as failure to give possession, but it may amount to much the same in practice. Strictly, the contractor is not entitled to an extension for the employer's failure to give possession (it is a serious breach of contract and grounds for a claim at common law), only for failure to give access.

5.1.2
ACA82

Clause 11 sets out your duties in dealing with extensions of time. They are summarised in flow chart 8.

After you have received a notice including all the particulars you consider necessary, you have a maximum of 84 days in which to consider and then award (if that is your decision) an extension of time. Obviously, you must act as quickly as possible.

If the contractor fails to give notice, together with the relevant information (see section 3.1.2) you have no obligation to issue any extension of time at all except as regards 'any act, instruction, default or omission of the Employer, or of the Architect on his behalf'. Your duty to award an extension in this situation may be exercised at any time before you issue the final certificate. Failure to carry out this duty would render liquidated damages unenforceable. Time would be at large and the employer would possibly take some action against you.

Within a reasonable time (probably not more than 84 days would be considered reasonable) after taking-over the works, you may, if you wish, review your previous decisions and either:

● confirm any date or dates for the taking-over of the works or any section *or*
● fix a later date for taking-over.

Note that you are not entitled to fix a date for

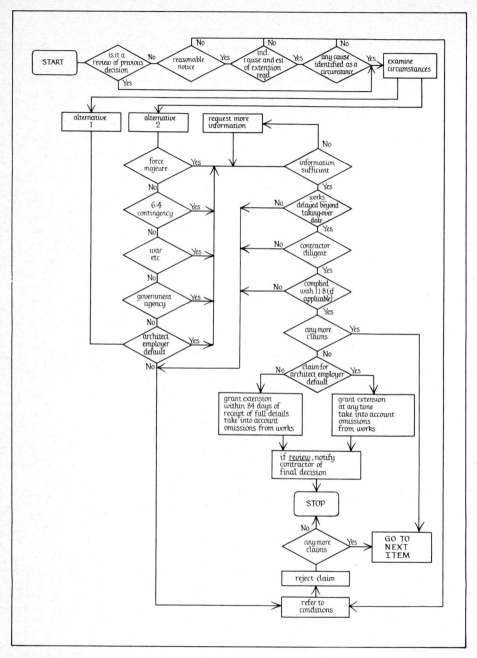

Flow chart 8   Architect's duties in relation to a claim for extension of time under ACA82

taking-over which is earlier than that previously fixed.

In order for you to carry out your review, the contractor is not required to submit further notices or information although he may well consider it wise to do so in order to receive the maximum benefit from your review. The review does not appear to be intended as an opportunity for you to consider new heads of claim, but merely to reconsider your previous decisions. However, occasionally it could be in the employer's interests to interpret the clause in a very broad way, depending upon particular circumstances.

Throughout your considerations, whether before or after taking-over, you are entitled to take into account the omission of work instructed since the previous extension was granted.

If the works are not fit and ready for taking-over on the date stated in the time schedule, you are bound to issue a certificate to that effect (clause 11.2) and the employer may instruct you to deduct liquidated damages or he may himself recover unliquidated damages (as the case may be) from the contractor. In the employer's interest, therefore, you must notify him (see letter 6) that further extensions may be due, otherwise he may have to refund the damages plus interest (clause 11.4).

Grounds for extension   It is appropriate, at this point, to consider the grounds for extension of time in clause 11 to decide what they mean in practical terms. It is vital that you clearly understand what may, and what may not, be grounds for extension.

The grounds included in clause 11.5 alternative 2 will be considered because clause 11.5(e) reproduces the ground in clause 11.5 alternative 1.

Alternative 1 relies upon your 'reasonable opinion'. Alternative 2 requires the contractor to prove to your 'satisfaction' that the grounds mentioned have prevented take over by the date or dates in the time schedule. A heavier burden is

6 Letter from architect to employer if work not completed by due date
but further extension may be due
This letter is suitable only for use with ACA82

Dear Sir,

I enclose my certificate in accordance with clause
11.2 of the contract.

You may, if you wish, recover liquidated damages
at the rate stated in the time schedule for the
period between the date the contract should have
been fit and ready for taking over and the actual
certified date of taking over. Alternatively, you
may instruct me to deduct the damages from the
next certificate.

Please bear in mind, when deciding whether or not
to deduct liquidated damages, that I have yet to
carry out my review of extensions of time. I
cannot begin until after taking over of the works
has occurred. Further extensions may be due. If,
as a consequence of my review, I fix a later date
for taking over, you would be liable to repay any
liquidated damages, together with interest,
deducted for the period up to the later date fixed
for taking-over.

Yours faithfully,

imposed on the contractor by alternative 2, but the precise difference may be difficult to identify. The best way to approach this particular problem is to ask yourself if you really believe that the taking-over of the works have been prevented by the grounds cited. Some people advise that you should try to look at each submission as though you were a complete stranger to the contract (i.e. you are not familiar with the particular works). This approach must be wrong, otherwise the contract would require the adjudicator to consider all applications for extensions of time. Instead, it specifically refers to the architect. You will have special knowledge of many of the circumstances, which it would be wrong, even if possible, to ignore.

CLAUSE 11.5(A): Force majeure

A difficult term to define because its meaning may alter depending upon the contract in which it is used. Many situations that would normally be covered by force majeure (such as war or riots) are already covered by other clauses. Consider this clause as a provision to cover events *not otherwise covered*, which are outside the control of the parties to the contract and which they did not contemplate when they entered into the contract. It would be reasonable to include a major strike or exceptionally adverse weather conditions in this category.

CLAUSE 11.5(B): The occurrence of one or more of the contingencies referred to in the insurance policy (clause 6.4)

This is straightforward. If the date of taking-over the works is delayed by any of the contingencies for which insurance has been taken out under clause 6.4, the contractor is entitled to an extension.

CLAUSE 11.5(c): War, hostilities, invasion, act of foreign enemies, rebellion, revolution, insurrection, military or usurped power, civil war, riot, commotion or disorder

Most of these grounds are self-explanatory. They range from the extreme (war) to the less serious (disorder). Thus anything more serious and including a disorder qualifies. A disorder would be a serious disturbance involving an element of violence.

CLAUSE 11.5(d); Delay or default by governmental agency, local authority or statutory undertaker in carrying out its statutory obligations in relation to the works

They must be statutory obligations and not any other type of work even if no other agency or authority can do it. The obligation must be in regard to the contract works, not to any other works even if delay is caused thereby. The clause covers such things as the laying of mains and cables to supply the works. It is reasonable to expect that the contractor will have included the work on his programme, showing its relation to the rest of site operations.

CLAUSE 11.5(e): Any act, instruction, default or omission of the employer, or of the architect on his behalf, whether authorised by or in breach of this agreement

This ground is the most important because if the contractor can show that the taking-over has been prevented by any act of the employer or architect and no extension has been granted, the date for completion becomes at large. The scope is extremely wide and covers anything done or not done by the employer or architect which causes the contract period to be exceeded. Some legal commentators have suggested that this ground does not cover anyone for whom the employer may be responsible (other than the architect). You would be wise, however, to consider extensions under this ground for the acts of all agents, employees or contractors engaged by the employer outside the terms of this contract.

5.1.3
GC/Works/1

Clause 28 sets out your duties in dealing with extension of time. It is assumed that you will also carry out the duties allocated to the authority in this respect. (Unlike the JCT80 and ACA82 forms, it is for the authority to decide upon extensions of time under this form, see section 1.3.3.) The duties are summarised in flow chart 9.

After you have received a notice from the contractor specifying the circumstances and other particulars, you must consider and then award (subject to the authority's approval) an extension of time. There is no time limit set in the contract so you must reach your decision within a reasonable time having regard to all the circumstances. Among the circumstances will be:

● Whether the contractor has furnished all the information you require. You have the right to ask for more information. If the contractor refuses, relying upon giving you the bare minimum as required by the contract, you must proceed with your estimation based upon what evidence you have. The result may be that the contractor will be awarded less than he expected. It will be his own fault.

● Whether the circumstance or circumstances causing delay have finished or if they are continuing. If they are continuing, for example a long strike or severe weather conditions, it is reasonable to wait until the delay is over before awarding an extension, even if the award is made after contractual completion date. In these circumstances some architects award a so called 'interim extension' and then top it up as soon as the delay is over. Such a procedure would benefit the contractor because it would reassure him that extensions were forthcoming and it would also

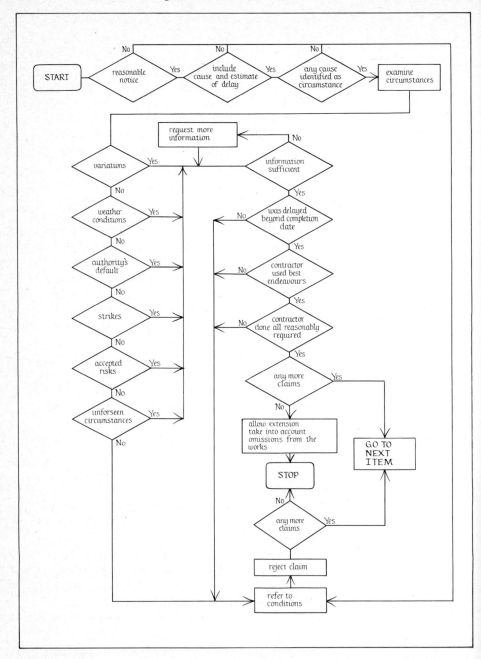

Flow chart 9   Architect's duties in relation to a claim for extension of time under GC/Works/1

show that you were alive to your responsibilities (see letter 3).

If any circumstance that you have to consider is under the control of the authority or yourself (i.e. clauses 28(2)(a) or (b)), you must act as soon as possible after you receive the contractor's notice in order to avoid the date for completion becoming at large. If the contractor fails to give notice as prescribed, the authority will almost certainly waive its right to notice in accordance with clause 28(2)(i) to enable you to act. In such a situation you should consider the award of an extension as soon as the circumstance comes to your attention but certainly before the contractual completion date.

There is no provision for a post-completion review in this contract. You can carry out such a review if the authority agrees. A notice by the contractor after completion of the works would not be acceptable because it would not be 'immediately upon becoming aware'. You (acting with the consent of the authority) are not allowed to reduce the amount of any previously fixed extension. Throughout your consideration, you are entitled to take into account the omission of work instructed since the previous extension was granted.

Clause 29(3) provides that the contractor is liable to pay liquidated damages if at any time the authority gives him notice that *in the opinion of the authority* the contractor is not entitled to any, or any further, extension of time. The clause provides for this procedure to be operated by the authority and you should be sure that the authority, and not you, issues the notice (even if you work within the authority) unless you have full powers to act in the name of the authority. Failure to operate the clause properly may cause problems in obtaining liquidated damages. It is suggested that, when such a notice is sent to the contractor, reference be made to the clause (see letter 7) to distinguish it

7   Letter from authority to contractor if contractor not entitled to any further extension
This letter is suitable only for use with GC/Works/1

Dear Sir,

I/We hereby give you notice under clause 29(3) of the contract that in my/our opinion you are not entitled to any/any further extension of time.

Yours faithfully,

7

from other letters that may be sent earlier rejecting claims for extension. This notice is clearly intended to be issued to the contractor after all extensions (if any) have been awarded.

Circumstances   It is appropriate at this point to consider the circumstances in clause 28(2) in order to decide what they mean in practical terms. It is essential that you clearly understand what may, and what may not, be grounds for extension.

CLAUSE 28(2)(A): The execution of any modified or additional work

If the delay is due to modified (i.e. altered or reduced but not omitted entirely) or additional work properly instructed, it will qualify for consideration under this circumstance.

CLAUSE 28(2)(B): Weather conditions that make continuance of work impracticable

Note that this does not mean the same as exceptionally adverse weather conditions in the JCT80 form. It does not matter that the contractor might have foreseen the possibility. A week of heavy snow in January would qualify, provided it was enough to stop the work. The only qualification is the overall proviso in clause 28(2)(iv) that the contractor must use his best endeavours to prevent or minimise the delay.

CLAUSE 28(2)(C): Any act or default of the authority

This circumstance is extremely important because failure to grant an extension would result in the date for completion becoming at large. The scope is wide and covers anything done or not done by the authority which causes delay. You should include in this circumstance any agents, employees or directly employed contractors.

CLAUSE 28(2)(D): Strikes or lockouts of workpeople

77

employed in any of the building, civil engineering or analogous trades in the district in which the works are being executed or employed elsewhere in the preparation or manufacture of things for incorporation

Note that two kinds of situation are not included—a strike affecting a trade, which affects a trade employed on the work (unless it be a manufacturing or preparatory trade) and a strike affecting transportation.

CLAUSE 28(2)(E): Any of the accepted risks

These risks are listed in clause 1(2) and they should pose little problem in interpretation (fire, tempest, floods etc).

CLAUSE 28(2)(F): Any other circumstance wholly beyond the control of the contractor

This circumstance is made subject to clause 28(2)(ii). No extension is possible if the contractor could reasonably be expected to have foreseen the circumstance *at the date of the contract* (not at the date of tender). The test is, therefore, whether the circumstance:

● is *wholly* beyond the control of the contractor *or*
● was not reasonably foreseeable at the date of the contract.

Such things as failure to obtain labour or materials might or might not qualify, depending upon the particular details of the case.

5.1.4
Agreement for Minor Building Works

Clause 2.2 states your duty quite simply. On receipt of the contractor's notification, you must make reasonable extension of time in writing.

There is no recital of relevant events. You must give an extension if the completion date will be exceeded for reasons beyond the control of the contractor. They need not be *wholly* outside the contractor's control. Such things as inability to obtain labour or materials must be judged on the merits of each case. If you or the employer cause delay, the contractor must be given an extension. Such things as severe weather conditions, strikes and statutory authorities' works must also be considered.

You must award the extension within a reasonable time. Because the contract period may well be very short, it will be impractical often to award an extension until after practical completion. The remarks in section 5.1.3 on this subject are generally applicable.

5.2 Preliminary matters

The technique for awarding an extension can be divided into several stages. After you have received the contractor's notice, estimates and particulars, you should read through them and decide whether, assuming that all the contractor says is true, there is a *prima facie* case for an award. This means that you must decide:

● If the contractor has correctly identified relevant events according to the contract.
● If he has included sufficient material for you to begin a detailed consideration.

Note that, at this stage, you are not concerned with proving or disproving the contractor's statements. For example, if the contractor claims under clause 25.4.2 of the JCT80 form, estimates 2 weeks for exceptionally adverse weather conditions and includes evidence in support, you will conclude that there is a case to consider. If, on the other hand, the contractor claims under a clause that is inappropriate, or his evidence does not suggest a valid claim, you will write and inform him without delay (see letter 8).

5.3 Collating

There are three basic techniques that can be used in collating the material to be considered. They can be summarised as:

● Reading and noting.
● Scheduling.
● The network base.

8 Letter from architect to contractor if claim for extension of time is
not valid

Dear Sir,

I have examined your claim for extension of time
which I received on the [*insert date*].

On the basis of the documents you have presented
to me, I see no grounds for any extension of time.
I shall be pleased to consider any further
submissions presented in the proper form and in
accordance with the contractual terms.

Yours faithfully,

They are not equally effective and they probably appeal to different types of architect. Whatever technique you choose, the first task is to put on one side any information that the contractor has submitted that does not appear to be strictly relevant. To this pile you will add any notes you make or other information you find that relates to the apparently irrelevant information. This pile may become relevant as your investigations proceed. At this stage, however, you have no way of knowing.

Reading and noting The first technique (reading and noting) is probably the most commonly used. It is included here for completeness and to indicate its drawbacks. It appears to be the easy way to collate material but, as so often is the case, the apparently easy way can lead to difficulties.

The technique involves taking each part of the contractor's claim in turn and checking it against readily available evidence (e.g. clerk of works' reports). At the end of the checking process you will have a series of ticks, crosses and queries against each statement in the contractor's claim. The queries must be resolved by looking for additional evidence from the contractor himself or from your own sources (see chapter 4). If a query cannot be resolved, despite all your efforts, you should put a cross against the item and ignore it. It is really the contractor's job to convince you.

If the contractor has put his claim together in a logical form, providing plenty of information, you will be in a position to make some sort of estimate. If the claim is at all complicated, inconsistent or confused, your task will be exceedingly difficult. The process is very much subjective and depends largely on your capacity to extract relevant information and keep a clear picture of the contract in your head. There is very little scope for you to take account of all the effects of delay, particularly the knock-on effects on site of which the contractor will be aware but which he may find very difficult to explain and prove adequately.

To summarise, this technique is adequate only if the claim is:

- simple
- presented clearly and
- backed by sufficient evidence.

Scheduling   The second technique (scheduling) is much more precise but has the disadvantage of requiring an abundance of patience and time. It does, however, produce a folder of information in tabulated form, which could well be useful if you are asked to explain your decision to an arbitrator.

The method involves setting down the contractor's claim in one column of the table in chronological order. In the next column briefly note the main points of his supporting evidence. Then gather all your own information pertinent to the various points of claim and jot it down concisely in the next column. It will be found easier to follow the claim through the contract if you stick all the sheets together vertically and fold them concertina-fashion. After all your own information has been added, carefully read each point of claim in turn and refer across to the evidence provided by the contractor and the evidence you have inserted. A number of gaps may become evident. You must try to resolve them by reference to the contractor and other members of the design team.

Table 6 shows a brief example of such a schedule prepared for a simple claim based upon the JCT80 standard form. In practice, you may well note evidence from a large number of sources.

At this stage, you will be able to complete the remarks column and form some idea of the delay caused by each event. The process produces a reasonably concise document containing all the available facts at your disposal. The secret of producing a good and useful schedule is to keep your notes clear but very brief so that it is possible to scan them quickly. It would be tempting but

Table 6   Scheduling (based on JCT80 SF)

Contractor's			Architect's		
Estimate of delay	*Claim*	*Supporting evidence*	*Evidence*	*Remarks*	*Estimate of delay*
10 days	Dated 14.11.84, Divert elec. cable caused delay 10 days behind programme at day 30 & except adverse weather in founds. wk.	Letter con. To authority placing order 3.9.84. Copy authority's letter stating 3 weeks notice necessary.	C/W report: work completed 12.10.84. Site meeting 3.10.84. con. said 'Not expected to hold up job'! Archt's. site visit 13.10.84 noted not ready for hardcore yet.	Notification somewhat late. Programmed to complete by 10.10.84.	Nil.
			C/W report: site scrape delayed 2 days due to plant breakdown.	Contractor's fault.	Nil.
			C/W report: take out old foundation delayed 6 days due to poor organisation on site.	Contractor's fault.	6 days no extension due.
			C/W report: mass conc. fill delayed 4 days due to 2 day's heavy rain & poor organisation on site.	Not except. adverse weather for October.	4 days no extension due.

Claim rejected.

untrue to say that at this point it is simply a matter of mathematics to determine the amount of extension. It must be borne in mind, however, that the amount of delay occurring during a contract is totally irrelevant unless it appears that the contractual completion date will be exceeded Two problems remain:

- Estimating the likely amount of overrun.
- Estimating how much of the overrun lies outside the control of the contractor.

With this method, you have the job of examining each portion of your synopsis and arriving at an

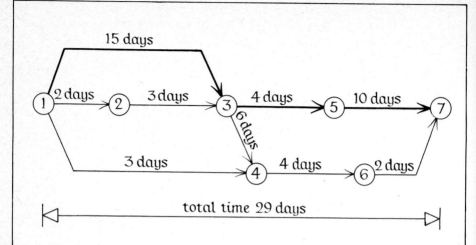

Figure 1   Effect of delay shown by a network analysis

To keep the example simple:

- Duration times are noted on each activity.
- Node points are numbered to identify each activity.
- Earliest and latest start and finish times are omitted.
- Earliest and latest event times are omitted.
- Critical path is shown as a thick black line.

Only delays in activities 1.3, 3.5 or 5.7 will affect the overall time unless delays in the other activities exceed the 'float' time. For example, if activity 1.3 is delayed by 2 days, it will add 2 days onto the total 29 days. But activity 1.2 could be delayed by up to 10 days without it affecting the overall total unless activity 2.3 is also delayed. If activity 2.3 is also delayed, the critical path becomes 1.2.3.5.7.

It is clear that the contractor could suffer considerable disruption (for which he may well have a financial claim) without any extension of time being appropriate.

intuitive amount of extension. The great advantage of this method over the first technique is that you have a useful worksheet to back up your final decision.

Network base   The third technique (the network base) is much superior to the previous techniques. It can be used whenever the contractor submits a network analysis as part of his evidence. It is always a good idea to require the contractor's master programme to be in the form of a network analysis. It can be done by inserting a suitable clause in the bills of quantities. The effect of the various events noted by the contractor can be seen immediately (see figure 1).

A network analysis for a building contract will occupy a sizeable piece of paper, depending upon the degree of detail. The technique for dealing with claims using the network is to write all the information, which you might otherwise put in schedule form, on the network itself in the form of brief notes at the relevant points of the network. Figure 2 shows a portion of network at the beginning of a contract. The claim is the same one shown in schedule form in table 6 for comparison purposes. The contractor has claimed an extension of time based upon delay by statutory authorities in diverting a cable. The notes are indicated briefly. In practice, you would take care to show the source of all information for future reference. By referring to earliest and latest event times, the effect of a delay can be seen. For example, it may be obvious that, although the available evidence points to a delay outside the contractor's control in one section of the network, it has no effect ultimately upon the critical path. In the example shown, the overall delay is caused by a series of delays of the contractor's own making. Although they occur in non-critical activities, the length of the delays make these activities critical. Important points to note are that:

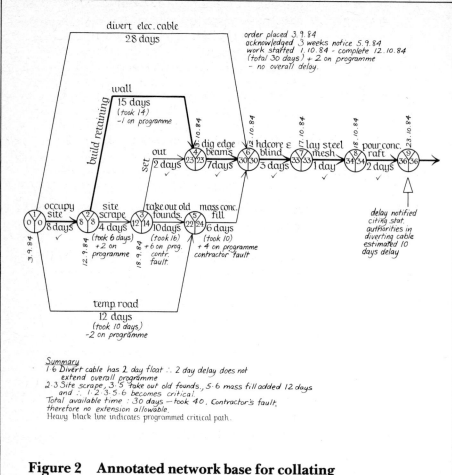

Figure 2   Annotated network base for collating

● The technique may be very time-consuming.
● A very clear and accurate picture of likely overrun is produced.
● The contribution of each delay can be determined without serious question.
● The effect of omissions can be seen.
● Dates, such as the latest date by which the contractor should have received information, can be established.

The process can be speeded up considerably if you

can put the network on a computer and immediately test the effect of delays to various portions of the network by feeding in the appropriate data. Comparison with the same information in bar chart form (see figure 3) clearly shows the advantages of a network to illustrate the real causes of a delay.

**5.4
Estimating the
extension of time**

Whatever method is chosen to collate the information, you will inevitably come to the point where you have to decide upon an extension of time. Two types of delay have to be considered:

● Concurrent (i.e. running at the same time).
● Consecutive (i.e. running one after the other).

Where two delays are concurrent and one clearly gives the contractor grounds for a financial claim (e.g. late instructions) while the other merely gives grounds for extension of time (e.g. exceptionally adverse weather) you may be wise to give the extension on grounds that do not imply financial consequences. It is still open to the contractor to try to prove loss as part of his financial claim. Consecutive delays leave no room for manoeuvre.

Confusion and doubt are often caused by delays, which—but for earlier delays—would not have occurred. The problem is twofold:

● Is the second delay to be taken into account?
● Should the grounds for extension for the second delay be the same as for the first or should it be considered on quite separate grounds?

A simple example based on the JCT80 standard form will explain the principle. Let us assume that your collation of information shows that the contractor has been delayed by 3 weeks on a critical activity due to delay on the part of a nominated supplier. Assume that special roof trusses are 3 weeks late and the roof covering is delayed. Before the roof covering can be completed, a week of severe frost is experienced

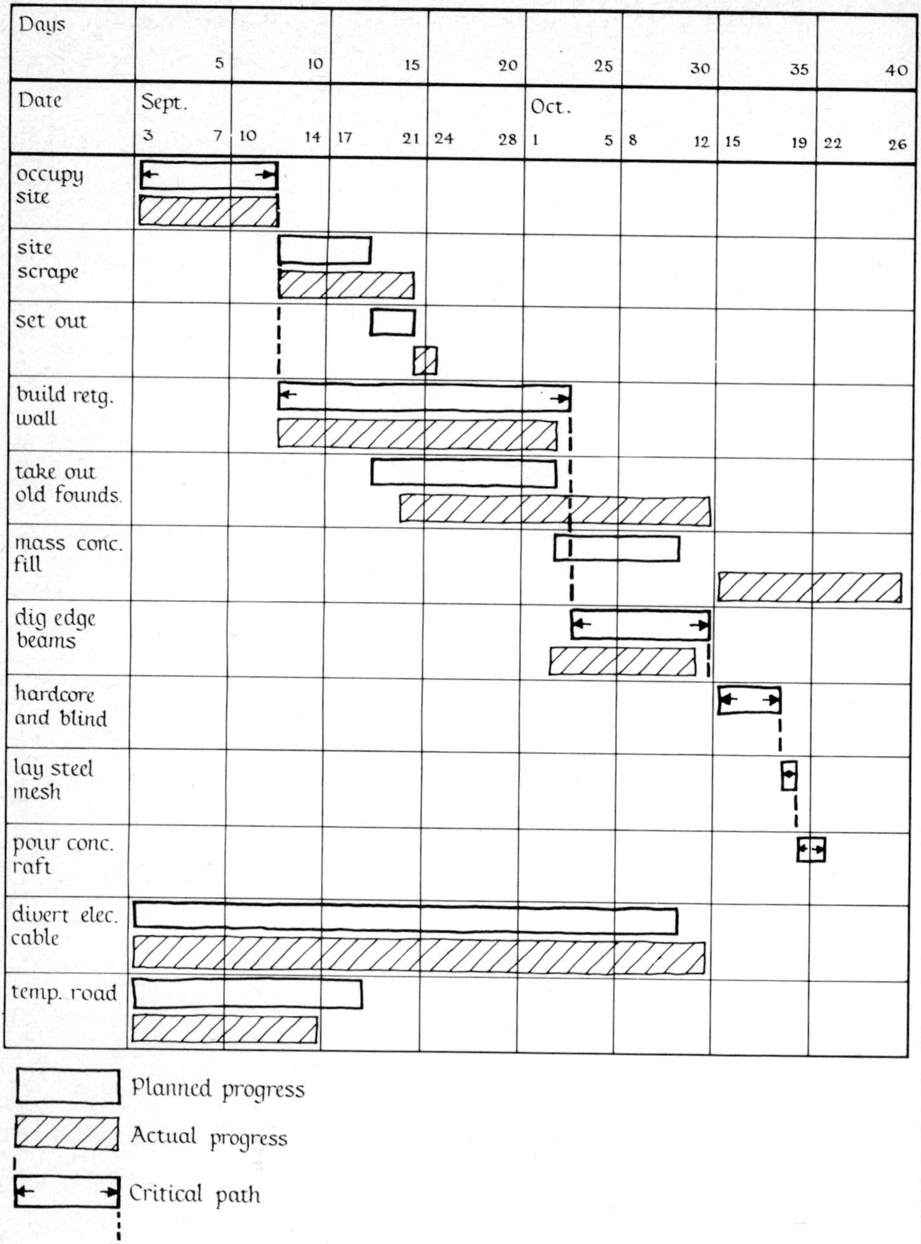

Figure 3 Bar chart

during which it is dangerous to work on the roof. A total of 4 weeks delay to internal work is therefore experienced. To what extension is the contractor entitled? It is clear that he should get 3 weeks extension under clause 25.4.7 for the delay on the part of the nominated supplier of special roof trusses, assuming all other conditions have been met, evidence is satisfactory and the contractor is unable to proceed with critical activities. The 1 week delay due to frost poses a different problem. Perhaps it occurred during February when frost is not particularly uncommon. Had it not been for the 3 week delay, the frost would not have merited consideration since it would not have ranked as 'exceptionally adverse weather conditions'. On the other hand, but for the 3 week delay, the frost would have caused no problem because, by that time, the building would have been covered.

You may be tempted to say that it is just the contractor's bad luck, as the frost could have occurred at any time. To take that attitude would be wrong. You are not required to speculate on what might have happened, but to consider what did happen. In this case it is clear that the delay due to the nominated supplier was not 3 weeks (i.e. the period between the date the trusses should have been ready and the date they were ready) but 4 weeks.

Delay 'on the part of the nominated supplier' was 3 weeks. On a restricted reading of clause 25.4.7, that is all you can award. Clause 25.3.1.1, however, refers to relevant events being the *cause* of the delay. Therefore, delay on the part of the nominated supplier amounted to 3 weeks and *caused* a further delay of 1 week. The whole extension of 4 weeks is due under clause 25.4.7. The example highlights the importance of reading clause 25 as a whole. A considerable degree of understanding of the building process is required to arrive at a fair and reasonable estimate of the extension of time in each case.

Contractors often complain that architects do

not properly understand what can be called the 'stone-in-the-pool' effect. A delay in one activity can spread ripples of delay throughout other activities which may not appear to be related. A network can show the effect to some extent, but often inexplicable delays in the critical path can be attributed to the ripple. The problem is that expected work can be suddenly removed from, say, a gang of bricklayers and the site agent's efforts to keep them occupied may upset careful programming. It is a familiar problem to contractors but it appears to cause architects some trouble.

The estimation of extensions must be carefully carried out in accordance with the appropriate contract that is being used. The rules must be strictly applied to suit the infinite variety of circumstances that can arise. In order to produce a fair and reasonable estimate, therefore, three things are required:

● Meticulous collation of information.
● Careful interpretation of the contract clauses.
● Thorough understanding of the nature of construction work.

5.5
Notifying the award

After you have arrived at your decision, you must inform the contractor immediately. Limit the information in your notification to:

● A list of relevant events.
● The appropriate extension for each event.
● Fixing the new completion date.

Do not be tempted to elaborate on your reasons for extension. It can lead to fruitless argument. That is not to say that you should not be willing to consider new evidence, but remember that once you have made your award you cannot alter it. You can, of course, make further awards as appropriate in accordance with the terms of the particular contract.

6 Techniques for dealing with loss and/or expense

It was noted in chapter 3 that the contractor has certain duties under the contract if he expects his claim to receive consideration. If the contractor does not observe those duties scrupulously you must reject the claim out of hand (see letter 9). However, the contractor still has his common law rights and if, despite your rejection, he presses you to consider his claim, you must refer the matter to the employer (the authority in the case of GC/Works/1) for his decision on the matter (see letter 10). The employer may well decide to waive his strict rights under the contract in order to secure a settlement without the necessity of litigation. Be sure to obtain his waiver in writing in unequivocal terms, which should authorise you expressly to deal with the matter.

6.1 Introduction

6.1.1 JCT80 Matters

The principal clause dealing with claims for loss and/or expense is clause 26. There are two other: clauses 13.5.6 and 34.3. Your duties in respect of clause 26 are set out in flow chart 10.

Assuming that the contractor has claimed in the correct form and at the proper time, or the employer has agreed to entertain a claim, you must consider the 'matters' that the contractor alleges have affected his progress in a substantial way. They are listed below.

CLAUSE 26.2.1: The contractor not having received at the proper time, necessary instructions etc. for which he specifically applied neither too early nor too late

You may be in breach of clause 5.4 but, unless the contractor applies in writing as noted, he has no

9   Letter from architect to contractor if claim rejected because of
non-compliance with contract conditions
This letter is not suitable for use with the Agreement for Minor Building Works

Dear Sir,

I refer to your claim for loss and/or
expense[1] received on the [*insert date*].

I cannot consider your claim because you have not
complied with your duties under the contract
clause 26[2]. In particular [*describe precisely,
but briefly, how the contractor has failed in his
duty*].

Yours faithfully,

1.   *Substitute 'damage, loss and/or expense' on
     ACA82.   Substitute 'expense' on GC/Works/1.*
2.   *Substitute '7' or '17' on ACA82.   Substitute
     '9' or '53' on GC/Works/1.*

10   Letter from architect to employer if the contractor presses for his claim to be considered
This letter is not suitable for use with the Agreement for Minor Building Works

Dear Sir,

I have received a financial claim from the
contractor. It was not submitted strictly in
accordance with the contract provisions and,
therefore, it has been rejected. However, the
contractor is pressing me to consider it. If I
continue to refuse, I suspect that he may exercise
his right to bring a claim at common law.

You may consider it advisable to waive your strict
contractual rights, in order to avoid litigation,
and instruct me to consider the claim as though it
had been submitted in accordance with the
contract.

I should be pleased to receive your decision as
soon as possible. If you require further advice,
please do not hesitate to telephone me.

Yours faithfully,

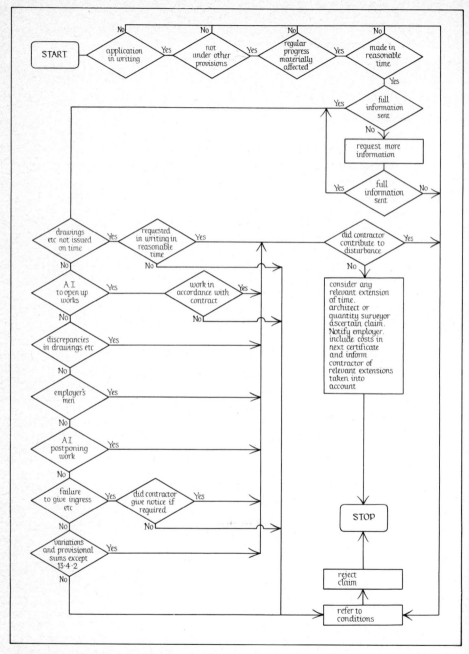

Flow chart 10   Architect's duties in relation to a claim for loss and/or expense under JCT80

contractual claim (see also the comments in section 5.1.1).

CLAUSE 26.2.2: Opening up for inspection or testing of any work or materials in accordance with clause 8.3 (including making good) found to be in accordance with the contract

This matter should be simply a question of fact.

CLAUSE 26.2.3: Any discrepancy or divergence between the contract drawings and/or the contract bills

If the contractor fails to notify you in time to allow an instruction to be given, he forfeits his right under this clause (which should be read with clauses 2.1 and 2.3). In fact he, not you, would be in breach of contract.

CLAUSE 26.2.4: The execution of, or failure by the employer to execute, work not forming part of the contract in accordance with clause 29 or the supply or failure by the employer to supply materials or goods which he has agreed to supply

This is often a source of severe disruption to the contractor because the employer perhaps fails to appreciate the situation and the onerous burden upon him (see also the comments in section 5.1.1).

CLAUSE 26.2.5: Architect's Instructions under clause 23.2 in regard to postponement

Included in this matter are instructions that may be construed as inevitably leading to postponement even if you do not refer to clause 23.2. An instruction directly to postpone is simple to identify.

CLAUSE 26.2.6: Failure of the employer to give, at the appropriate time, ingress to or egress from the site or any part including necessary passage over adjoining land in the possession of the employer after the contractor has given any notice required by the contract documents or failure of the employer to give ingress or egress as agreed between the architect and contractor

See comments in section 5.1.1.

CLAUSE 26.2.7: Architect's Instructions under clause 13.2 (variations) or clause 13.3 (provisional sums) except where the contractor tenders for the work

This refers to any loss and/or expense which is additional to the cost of the instruction itself or the work covered in the provisional sum. For example, if you instruct the contractor to add showers in each bathroom on a large housing scheme, he will be paid the appropriate amount for the additional materials and work involved under clause 13, but he may also incur expense because the operatives have to be diverted from other work.

Two other clauses contain grounds for loss and/or expense and are listed below.

CLAUSE 13.5.6: Valuations not relating to additions, substitutions or omissions or liabilities dealt with under clauses 13.1–13.5, not reimbursable under any other provision of the contract.

See comments in section 1.5.1.

CLAUSE 34.3.1: Discovery of antiquities

See comments in section 1.5.1.

6.1.2
ACA82 The principal clause dealing with claims for damage, loss and/or expense is clause 7. There are two others: clause 10.3 (included by reference in clause 7.2) and clause 17. Your duties in respect of clauses 7 and 17 are set out in flow chart 11.

Events Assuming that the contractor has claimed in the correct form and at the proper time, or the

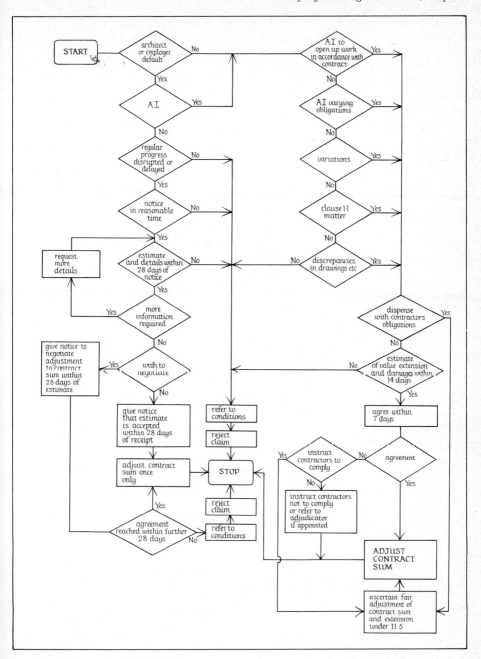

Flow chart 11   Architect's duties in relation to a claim for damage, loss and/or expense under ACA82

employer has agreed to entertain a claim, you must consider the events which the contractor alleges have affected his progress in a substantial way. They are listed below.

CLAUSE 7.1: Any act, omission, default or negligence of the employer or of the architect (except Architect's Instructions)

The scope is extremely wide. It means exactly what it says. You must consider under this head any claim relating to anything done or not done by the employer or yourself (except Architect's Instructions). The clause is broad enough to include claims which would normally be considered ex-contractual under the JCT80 form. For example, if you fail to issue a certificate in accordance with clause 16.2, the contractor could make a claim under this event and you would have a duty to consider it.

CLAUSE 10.3: If the execution of any work or installation of materials etc, not forming part of this contract, by the employer or his agents disrupts the regular progress of the works or delays execution in accordance with the time schedule and the contractor thereby incurs damage, loss and/or expense

Work carried out by the employer can cause the contractor severe disruption. Included is work carried out by statutory authorities not in pursuance of statutory obligations and paid direct by the employer.

CLAUSE 17.1: Instructions issued in relation to opening up or testing of work or materials, addition, alteration or omission of obligations or restrictions regarding limitation of working space, hours, access or use of site, alteration or modification of design, quality or quantity of works etc, plus any other matter, any ambiguity or discrepancy in contract documents and statutory requirements

The opening up or testing must have resulted in the work or materials being found to be in accordance with the contract for a claim to be valid. The ambiguity or discrepancy must not be such that they could have been found or were reasonably forseeable by the contractor at the date of the contract. The other provisions are straightforward. Note that if the instruction requires work provided for or reasonably to be inferred from the contract documents, the claim is not valid.

6.1.3
GC/Works/1

The principal clause dealing with claims for expense is clause 53. There is also clause 9(2)(a)(i). Your duties in respect of clauses 53 and 9(2) are set out in flow chart 12.

Circumstances

Assuming that the contractor has claimed in the correct form and at the proper time, or the authority has agreed to entertain a claim, you must consider the circumstances which the contractor alleges have affected his progress in a substantial way. They are listed below.

CLAUSE 53(1)(A): Complying with any Architect's Instructions given or confirmed in writing

This clause refers to any expense additional to the cost of the instruction itself.

CLAUSE 53(1)(B): The making good of loss or damage caused to the works by neglect or default of a servant of the Crown acting in the course of his employment or by any of the accepted risks

This clause also refers to disruption expenses additional to the actual cost of making good (which will be valued as a variation) although, in practice, it may be difficult to separate the two aspects. In any case, you have to decide whether the clause applies before the contractor is due to any payment at all. Note that expense caused by the action of a servant of the Crown *not* acting in

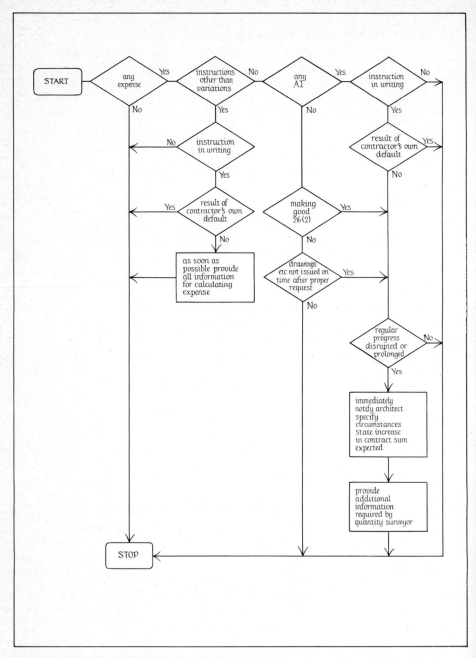

Flow chart 12   Architect's duties in relation to a claim for expense under GC/Works/1

the normal course of his employment would not lead to a claim against the authority. If the expense was due to a cause outside the two categories (servant of the Crown or accepted risks) there would be no valid claim even though loss and damage was caused which had to be made good. The words 'neglect or default' should be taken to mean either:

- doing that which ought not to be done *or*
- not doing that which ought to be done.

CLAUSE 53(1)(C): The execution of work on site by persons engaged directly by the authority

This can be a source of severe disruption. Included is work carried out by statutory authorities not in pursuance of their statutory powers and paid for directly by the authority. Note that this clause would cover work being carried out to another building provided that it is on the same site and the work causes disruption to the contractor's progress.

CLAUSE 53(1)(D): Delay in the provision of drawings, schedules, levels or other design information (unless to be prepared by the contractor or any of his sub-contractors), delay in the execution of work or provision of materials by the authority (unless due to the fault of the contractor), delay in nomination or dealing with prime cost sums provided that the contractor shall have given you notice in reasonable time specifying the item and date required or you have otherwise agreed with him a date for provision of the item

This clause is basically very straightforward and deals with delays in providing information or services. You should have no difficulty in deciding if there has been a delay but, remember, the contractor must have applied in writing, neither too early nor too late, in order to have a valid claim. Your judgement must take into account your own as well as the contractor's convenience.

CLAUSE 9(2)(a)(i): Complying with Architect's Instructions (other than alterations, additions or omissions from the works), provided that the expense is not otherwise provided for or reasonably contemplated by the contract

This clause covers the situation where you have issued an instruction in regard to discrepancies in the contract documents, the opening up for inspection of any work found to be in accordance with the contract and any other instruction not being a variation to the works.

6.2
Preliminary matters

When the claim is received, you must read it carefully and decide whether the contractor has a *prima facie* case. He will have a case worth considering if he has:

● Correctly identified the appropriate matters according to the contract.
● Submitted sufficient particulars for you to begin a detailed consideration.

You should not concern yourself with the value of the contractor's claim unless you intend to ascertain the amount of loss and/or expense yourself. The contract allows you to ascertain the amount of loss and/or expense or to instruct the quantity surveyor to do so. You are strongly advised to instruct the quantity surveyor to ascertain the amount. He is the building economist and better suited to the task even if the amount is relatively small. In any case, the value is of no consequence at this stage because the claim may not be valid. It is your responsibility to decide validity. You cannot delegate that job to the quantity surveyor or to anyone else.

If you come to the conclusion, at this stage, that the claim is not valid either because it is obviously just a 'try on' or because the evidence submitted does not support it, you must inform the contractor immediately (see letter 9).

6.3
Collating

The techniques for collating the information in respect of a financial claim are basically the same as those you use in considering a claim for extension of time (see section 5.3). Contractors differ in their ability to submit clear and precise claims. Before attempting to produce a collation, you must identify all ex-contractual and *ex-gratia* claims and put them to one side to be referred to the employer.

6.4
Deciding validity

In carrying out your duty under the contract, remember that it consists of only one basic decision: whether or not the claim made by the contractor is valid. Many side issues can conspire to confuse the question, but during the process of collation you should have removed them.

You must take the claim, or—if the claim is in a number of parts—each part separately, and consider it in relation to the information you have set out in your collation. In each case ask yourself the following questions:

● Does the claim fall within the scope of one of the matters referred to (in clause 26.2 of JCT80, clause 7 or clause 17 of ACA82 or clause 53 or clause 9 of GC/Works/1)?
● Has the regular progress of the work been affected by the matter, whether or not the overall completion date has been or is likely to be exceeded?
● Has the progress been affected substantially?

If the answer to all your questions is 'Yes', the claim is valid and you should pass your opinion to the quantity surveyor for ascertainment (see letter 11) and to the contractor for his information. If you decide that the claim is partially valid, that is to say you can answer 'Yes' to all the above questions in respect of part of the claim, make the position clear to the quantity surveyor (see letter

12) and the contractor (see letter 13).

6.4.1
JCT80

Note that you must inform the contractor what extension of time, if any, has been made under clause 25 which is to be taken into account in ascertaining loss and/or expense under clause 26. The relevant events can be only one or more of the following:

- clause 25.4.5.1 (referring to clauses 2.3, 13.2, 13.3 and 23.2)
- clause 25.4.5.2
- clause 25.4.6
- clause 25.4.8
- clause 25.4.12.

Remember that none of these clauses may be appropriate.

The process of ascertainment must not be delayed after you have agreed the validity of the claim. If the full amount cannot be settled quickly by the quantity surveyor, you must request him to notify you as soon as he ascertains any part of the claim so that the amount can be included in the next certificate.

Clauses 13.5.6 and 34.3.1 do not strictly require the contractor to claim although he is almost certain to do so. In any case, you will require him to submit full details of any loss and/or expense so that you can properly carry out your duties. If he refuses to supply information in such cases, you must still carry out the work of collation and valuation to the best of your ability. The contractor will only have himself to blame if the amount eventually paid is lower than he expected or even nothing at all.

6.4.2
ACA82

Note that, under clause 7, there is a strict timetable laid down:

- The contractor has 28 days after giving notice of the event, to submit estimates etc.
- You then have 28 days, from receipt of

11   Letter from architect to quantity surveyor requesting him to
ascertain the amount of payment
This letter is not suitable for use with the Agreement for Minor Building Works

Dear Sir,

I enclose a claim dated [*insert date*] received
from the contractor in respect of loss and/or
expense.[1]

After careful consideration, my conclusion is that
the claim is valid. I should be pleased if you
would proceed as soon as possible to ascertain the
payment due to the contractor.

[*Add, if using ACA82*]
In order to comply with the contract conditions,
your ascertainment must be complete and in my
hands by [*insert date*] at the latest.

Yours faithfully,

1. *Substitute 'damage, loss and/or expense' on
   ACA82. Substitute 'expense' on GC/Works/1.*

12   Letter from architect to quantity surveyor if part of the claim is
valid
This letter is not suitable for use with the Agreement for Minor Building Works

Dear Sir,

I enclose a claim dated [*insert date*] received
from the contractor in respect of loss and/or
expense.[1]

After careful consideration, I conclude that part
of the claim is valid. [*State briefly and
precisely, referring to the points of claim, which
items are valid, which items are partially valid
and in what respect, which items are invalid*].

I should be pleased if you would proceed as soon
as possible to ascertain the payment due to the
contractor.

[*Add, if using ACA82*]
In order to comply with the contract conditions,
your ascertainment must be complete and in my
hands by [*insert date*] at the latest.

Yours faithfully,

1. *Substitute 'damage, loss and/or expense' on
   ACA82. Substitute 'expense' on GC/Works/1.*

13   Letter from architect to contractor if part or whole of the claim is valid
This letter is not suitable for use with the Agreement for Minor Building Works

Dear Sir,

I refer to your claim for loss and/or expense[1] dated [insert date]. After careful consideration of the evidence, I am of the opinion that there is some merit in your claim and I am asking the quantity surveyor to ascertain the amount due to you.

[Add either]
The amount so ascertained will be added to the next certificate after ascertainment has been completed.

[Or (If ACA82 clause 17.1 used)]
I will inform you of the amount so ascertained in order that you can notify me of your agreement in accordance with clause 17.2 of the contract.

[Or (If ACA82 clause 7 used)]
You are to take this letter as notice under clause 7.2 of the contract that I wish to negotiate any adjustment to the contract sum. To achieve this end, I should be pleased if you would co-operate with the quantity surveyor. When I receive his ascertainment, I will inform you so that you can notify me of your agreement.

Yours faithfully,

Copy: Quantity Surveyor

1. *Substitute 'damage, loss and/or expense' on ACA82. Substitute 'expense' on GC/Works/1.*

estimates etc, to give notice that you accept them or you wish to negotiate.

● You have a further 28 days, after the date of your notice, to reach agreement with the contractor.

In practice, you have 28 days to decide if the claim or any part is valid and another 28 days to ascertain the precise amount, if any, payable and agree it with the contractor. You would be wise to allow the quantity surveyor to deal with the financial aspect.

If agreement cannot be reached within the total period laid down, either the contractor or the employer may apply for adjudication if clause 7.4 is not deleted. The employer is unlikely to do so. The contractor is unlikely to do so either, provided that he can see some likelihood of reaching agreement on reasonably advantageous terms. If clause 7.4 has been deleted then you (or more likely the quantity surveyor) will have to continue negotiations until agreement is reached or until you decide that you must settle the matter formally in accordance with clause 25.1 (alternative 2) by unilaterally deciding the matter. The contractor or employer can, of course, seek arbitration, but it cannot take place until after taking-over of the works.

Unlike the JCT80 provisions, you are only allowed to make one adjustment of the contract sum in respect of any particular claim. If, therefore, the contractor submits further evidence of continuing disruption caused by an event for which you have already issued a certificate, you must not consider it no matter how good a case he may have. Write him a letter and explain (see letter 14).

Clause 17 lays down a stricter timetable:

● The contractor has 14 days (or such period as you may agree with him) from receipt of an Architect's Instruction, to submit estimates of the value of the instruction, any extension of time

14   Letter from architect to contractor if further evidence submitted
after contract sum adjusted under clause 7
This letter is suitable only for use with ACA82

Dear Sir,

Thank you for your letter of the [*insert date*]
with which you enclosed further information in
respect of your claim for damage, loss and/or
expense dated [*insert date*].

The contract sum has already been adjusted, to
cover the amount agreed for the claim, in my
certificate number [*insert number*] dated [*insert
date*]. The contract expressly states, at the end
of clause 7.2, that in these circumstances 'no
further or other additions shall be made in respect
of such claim'. Therefore, I return your further
information with this letter.

Yours faithfully,

claimed, damage, loss and/or expense.
● You have only 7 days, from receipt of the contractor's estimates etc, to agree them.

You are unlikely to be able to agree the contractor's estimates in 7 days. If you can, that is the end of the matter. If you cannot, you must either:

● instruct the contractor to comply nevertheless and ascertain the value of the instruction and any damage, loss and/or expense in accordance with clause 17.5; *or*
● instruct the contractor not to comply (he then has no claim (clause 17.4); *or*
● refer the estimates to the adjudicator, if any, for a decision.

You can dispense with the whole procedure if you wish (clause 17.5) by so informing the contractor before or after you issue your instruction (see letter 15). You must then decide upon validity and either you, or better still the quantity surveyor, will ascertain the payment to be made. Although the clause does not specifically state that the payment must be certified in one operation, you would be wise to do so, as in clause 7. You are not specifically prevented from considering a late submission by the contractor. However, you can decline to consider it if you have agreed the contractor's estimates. 'Agreed' must be taken to mean a negotiated and settled figure.

The comments on adjudication and arbitration given earlier in this chapter are equally applicable to clause 17.

6.4.3
GC/Works/1

Note that, under clauses 53 and 9(2), there is no time limit set for you to decide validity or the quantity surveyor to ascertain the amount of expense. Therefore, you must act as quickly as possible. Clause 40(5) makes it clear that sums added to the contract sum on account of clauses 53 or 9(2) must be added to sums to be paid on your

15 Letter from architect to contractor dispensing with contractor's
obligations under clause 17.5
This letter is suitable only for use with ACA82

Dear Sir,

[Either]
I enclose my Architect's Instruction number
[insert number] dated [insert date].

[Or]
I refer to my Architect's Instruction number
[insert number] dated [insert date].

[Then]
In this instance, I have decided to dispense with
your obligations under clause 17.1 and put clause
17.5 into operation. Therefore, you should comply
with the instruction immediately.

Yours faithfully,

Copy: Clerk of Works [if appointed]

financial certificates issued in accordance with clause 40(3). The implication is that such sums must be included in the next certificate after ascertainment.

6.5
Points to note

● The contractor should specify the clauses under which he is expecting payment. If he does not, ask him to do so.

● If the contractor expects payment, you have the right to ask for any information that will assist you to come to a decision.

● Do not give the contractor the benefit of the doubt. He must prove his case to your satisfaction.

● Report your findings to the employer (authority in the case of GC/Works/1). Do not leave him to query a payment included in the certificate. The employer, however, has no right to interfere with your decision.

7 Claims from sub-contractors

7.1
NSC/4 or NSC/4a
(used with JCT80)

7.1.1
Duties of parties:
extension of time

Under clause 11 of sub-contract NSC/4 or NSC/4a as appropriate, the contractor has a duty to inform you immediately he receives any written notice from a nominated sub-contractor in respect of delay and he must submit all written representations made by the nominated sub-contractor. If the nominated sub-contractor so requires, the contractor must join with him in requesting your consent (under clause 35.14 of the main contract) to an extension of time. He must give an extension of time to the nominated sub-contractor if you give your consent. With your agreement, he must:

● State which of the relevant events have been taken into account.
● State the extent to which you have had regard to any variation requiring omission of work since the previous fixing of a revised period.

The relevant events referred to in clause 11 of NSC/4 or NSC/4a closely parallel the relevant events in clause 25 of the main contract and they will not be commented on further. You should, however, carefully read them when considering the application and study the notes in section 5.1.1.

The contractor must fix the revised periods no later than 12 weeks from receipt of reasonably sufficient particulars and estimates or before the expiry of the previously fixed period for completion of the sub-contract works. After the first exercise of his duty, the contractor may, on subsequent occasions, fix an earlier date (with your consent).

Not later than 12 weeks after the practical

completion of the sub-contract works, the main contractor must review the situation and (with your consent) either:

● fix an earlier date for completion of the sub-contract works; *or*
● fix a later date for completion of the sub-contract works; *or*
● confirm the date previously fixed.

If the nominated sub-contractor fails to complete the work by the completion date, the main contractor must notify you and give the nominated sub-contractor a copy of the notice.

Your own duty arises under clauses 35.14 and 35.15 of the JCT80 standard form. Essentially, it is for the contractor to seek your *consent* to an extension of time. Therefore, the contractor should present the information together with his conclusions regarding the allowable extension. You must check the details carefully. It is a sensible idea to go through the process of collating described in section 5.3. If you are reasonably satisfied that the contractor has presented the information and drawn conclusion correctly, you must give your written consent. If you are not satisfied, it is not advisable for you to suggest a different period of extension, but merely to inform the contractor that you do not think the extension suggested is correct.

In practice, the contractor (and probably the nominated sub-contractor) will desire a meeting with you to discuss any points of difference. Remember at all times that, although your consent is required, it is up to the contractor to take the lead. You can stand firm until the contractor proposes an extension which accords with your own thoughts on the matter. But, also bear in mind that the contractor and nominated sub-contractor are bound to have a much more detailed knowledge than you of the sub-contract situation.

If you receive a notice from the contractor that

the nominated sub-contractor has failed to complete the sub-contract works by either:

● the period specified in the sub-contract; *or*
● any extended time granted by the contractor (with your consent),

you must write to the contractor (with a copy to the nominated sub-contractor) certifying the failure, provided that you are satisfied that the provisions of the extension of time clause have been properly carried out. If you are not satisfied, do not issue your certificate. The withholding of your certificate is a useful way of ensuring that the contractor has properly represented the nominated sub-contractor's case to you. Resist any temptation to deal directly with the nominated sub-contractor. His contract is with the main contractor through whom, unless there are very exceptional circumstances, all your information should pass.

7.1.2
Duties of parties:
loss and/or expense

The contractor has mandatory duties relating to claims for loss and/or expense by a nominated sub-contractor under clause 13 of sub-contract NSC/4 or NSC/4a referred to in clause 26.4 of the main contract. He must:

● Require you to operate clause 26.4 if he receives a written claim from the nominated sub-contractor in the proper form.
● Pass to you whatever information you may need to form an opinion.
● Pass to you whatever details you may need to carry out an ascertainment.

Your duty arises under clause 26.4 of the JCT80 standard form. When you are satisfied that you have received all the information that you require from the main contractor, you must proceed with the ascertainment as discussed in chapter 6 (so far as it relates to the JCT80 form). The matters referred to in clause 13 of NSC/4 or NSC/4a closely parallel the matters in clause 26 of the main contract. You should, however, carefully read them when considering the application and study the notes in section 6.1.1.

Note that you must inform the contractor (copy to nominated sub-contractor) to which extensions of time, if any, of the sub-contract work you have given your consent and which is to be taken into account in ascertaining loss and/or expense under sub-contract clause 13. The relevant events can only be one or more of the following from NSC/4 or NSC/4a:

● clause 11.2.5.5.1 (referring to clauses 2.3, 13.2, 13.3 and 23.2 of the main contract conditions)
● clause 11.2.5.5.2
● clause 11.2.5.6
● clause 11.2.5.8
● clause 11.2.5.12.

Remember, none of the clauses may be appropriate.

**7.2
ACA sub-contract
(used with ACA82)
and sub-contracts
used with
GC/Works/1**

You have no special duties in regard to extensions of time as far as sub-contractors are concerned. They are to be dealt with by the main contractor. You will only be concerned with damage, loss and/or expense insofar as the main contractor includes sub-contract items in his own claim.

**7.3
Agreement for
Minor Building
Works**

You have no special duties in regard to extensions of time as far as sub-contractors are concerned. They are to be dealt with by the main contractor. You have no power to consider claims for loss and/or expense either from a sub-contractor or main contractor (see section 10.5).

8 Liquidated damages, penalties and bonus clauses

**8.1
Liquidated and
unliquidated
damages**

Liquidated damages have been the subject of a great deal of misunderstanding among architects, quantity surveyors and contractors. It is worthwhile getting the position absolutely clear. Liquidated damages are a genuine pre-estimate of the damage that would be suffered if a certain event took place. It is a device adopted to avoid the necessity and trouble of proving actual damage after the event. It is used in the JCT80, ACA82, GC/Works/1 and the Agreement for Minor Building Works contracts to calculate the amount payable by the contractor if he exceeds the completion date or any extended date. The sum is normally expressed as a sum of money per day or per week of overrun.

If no sum was included in this way, the employer would not be able to claim damages for any loss and/or expense he may have suffered without proving the amount. Any damages so proved are said to be unliquidated. The advantages of liquidated damages are that:

● They do not have to be proved.
● They are agreed between the parties in the sense that they are known to the contractor at the time of tender and he can allow for them in his tender sum. (It is not unknown for a contractor to estimate how long he might overrun, multiply the number of weeks by the liquidated damages and allow for the resultant sum in his tender figure.)
● The employer can simply deduct them. He does not have to issue a writ through the courts.

It does not matter whether the liquidated damages are a true reflection of the actual damage suffered. They may be greater or less. If they were a realistic estimate of the damage envisaged at the time the contract was made, the employer can deduct them if the completion date is exceeded. The system can work against the employer because he does not have the option to issue a writ for unliquidated damages if the liquidated sum is seriously below the damage actually suffered. Once the sum is fixed, the employer must adhere to it.

8.2 Penalties

A penalty is either a predetermined sum which is not a realistic pre-estimate of damage, or a sum which is payable on the occurrence of any one of a number of different kinds of events. A penalty is not enforceable, but the employer is entitled to recover his actual loss in an action for common law damages. Whether the sum is described as a penalty or liquidated damages does not matter. The vital point is the true nature of the sum. Liquidated damages clauses are often referred to incorrectly as 'penalty clauses'.

8.3 Bonus clauses

It is not common to find a bonus clause in a contract but the employer may decide to include such a clause if he wishes the work to be completed as soon as possible. It is the reverse of liquidated damages, but the employer may nominate any amount he wishes to be paid to the contractor for every week or day, as the case may be, that the contractual completion date is anticipated. The bonus figure must be notified to the contractor at the time of tender so that it can be taken into account. There are two particular dangers in bonus clauses:

● The contractor will be more anxious than usual to finish quickly. He may skimp the work, and instructions may be challenged more frequently. Generally, disputes are more likely.
● The relationship between the amount in the bonus clause and the amount of any liquidated damages poses problems that are not easily resolved in view of the fact that the liquidated

damages, unlike the bonus clause, must be a genuine pre-estimate. The contractor may see the contrast as unfair, depending upon whether he finishes early or late, and a lengthy dispute can develop.

If your client is keen to include a bonus clause, you would be wise to suggest instead a shorter contract period with a realistic liquidated damages clause. A higher tender figure may result but probably no higher than the equivalent sum which would result from a lower tender figure (based on a longer period) plus the appropriate weeks' bonus. Moreover, you would certainly have less problems in running the contract. It may appear that a contractor working to a short programme would be just as liable to create problems because he would be under pressure. Certainly you would need to ensure that your own procedures were punctual, but the contractor would know that his tender figure was secure provided he finished on time. The difference is subtle but real. In one situation, the contractor is fighting to make a good profit, in the other, his profit is reasonably certain.

**8.4
Calculation of
liquidated damages**

Liquidated damages must be a genuine pre-estimate of future loss and/or expense. With this in mind, you must decide upon a suitable figure at the latest possible moment to allow insertion in the bills of quantities. You must agree the figure with the employer and confirm it in writing (see letter 16) so that there is no room for dispute later.

   In arriving at the appropriate sum, the employer will play a leading part because only he will know the likely effect of delay on his finances. However, he will rely upon you to point out the kind of expenses that are allowable. Among the categories of expense are the following:

● Rental of alternative and equal (but not better) premises.
● Storage costs of equipment and furniture.

16   Letter from architect to employer to agree the amount of liquidated damages

Dear Sir,

I refer to our conversation of the [*insert date*] and confirm that you have agreed a figure of £ [*insert amount*] per week to be included as liquidated and ascertained damages in the contract documents. The figure is worked out as follows:

[*List and total the weekly sums taken into account*]

Yours faithfully,

Copy: Quantity Surveyor

● Loss of anticipated production profits (this should be a conservative estimate).
● Additional removal expenses.
● Additional salary expenses.
● Additional professional fees.
● Any other expense, or loss suffered, which would not be incurred unless the contract period is exceeded.
● A percentage to cover inflation at the current rate.

Usually the sums should be totalled on a weekly basis. If calculation is difficult, the best possible estimate must be used. Besides your letter of confirmation, you should keep your calculation papers in a safe place, in case the sum is questioned during litigation or arbitration.

8.5
Deduction of
liquidated damages

The weekly sum is not subject to justification before deduction. The employer is entitled to deduct liquidated damages from the next financial certificate after you have issued a certificate stating that the contractor should have completed by a certain date (no certificate is required with GC/Works/1, see section 5.1.3). If insufficient money is due to the contractor to allow deduction of liquidated damages, they become a debt payable to the employer.

In making deductions, the employer must state why the deductions are being made and indicate how the sum is calculated (see letter 17).

If you grant a further extension after liquidated damages have been deducted and issue a further certificate stating a new date on which the works should have been completed, the employer must refund all money withheld on account of the second extension period. The contractor may be entitled to receive interest on the money incorrectly withheld. The ACA82 form specifically gives him this entitlement. It is wise to warn the employer if further extensions may be due so that

17 Letter from employer to contractor when deducting liquidated damages

RECORDED DELIVERY

Dear Sir,

I enclose my cheque in the sum of £ [*insert amount*] as payment due under certificate number [*insert date*] dated [*insert date*]. Liquidated damages have been deducted in accordance with clause 24[1] of the contract. They have been calculated as follows:

Contractual completion date: [*insert date or extended date*].

[*Then either*]
Practical completion[2] date: [*insert date*].

[*Or*]
Date at which liquidated damages calculated: [*insert date*].

[*Then*]
= [*insert number*] weeks @ £ [*insert amount*] per week = £ [*insert total*].

Yours faithfully,

1. *Substitute '11' on ACA82. Substitute '29' on GC/Works/1. Substitute '2' on Agreement for Minor Building Works.*
2. *Substitute 'taking over' on ACA82. Substitute 'completion' on GC/Works/1.*

18   Letter from architect to employer regarding the deduction of
liquidated damages

Dear Sir,

Thank you for your letter/I refer to your telephone
call [omit as appropriate] of the [insert date]
regarding the deduction of liquidated damages.

[Add if using JCT80 or Agreement for Minor Building
Works]
You are entitled to deduct liquidated damages at
the rate of £[insert amount] per week for every
full week between the date at which the contractor
should have completed as certified by me and the
date of practical completion.

[Add if using ACA82]
You are entitled to deduct liquidated damages at
the rate of £[insert amount] per week for every
full week between the date at which the works were
fit and ready for taking-over as certified by me
and the date of taking-over.

[Add if using GC/Works/1]
You are entitled to deduct liquidated damages at
the rate of £[insert amount] per week for every
full week between the date at which the contractor
should have completed and the date on which the
works were completed and the site cleared,
provided that you have notified the contractor
that, in your opinion, he is not entitled to any
or any further extension of time.

Yours faithfully,

19   Letter from architect to employer if advice required on whether to deduct liquidated damages

Dear Sir,

Thank you for your letter/I refer to your telephone call [*omit as appropriate*] of the [*insert date*].

You are entitled to deduct damages at the appropriate rate. The decision to deduct liquidated damages is a matter reserved by the contract for you alone. There may be considerations, unknown to me, which will influence your decision.

Yours faithfully,

19

20   Letter from architect to employer if it would be unfair to deduct liquidated damages

Dear Sir,

Thank you for your letter/I refer to your telephone call [*omit as appropriate*] of the [*insert date*].

You are entitled to deduct liquidated damages at the appropriate rate. The decision to deduct is a matter reserved by the contract for you alone.

There may be considerations, unknown to me, which will influence your decision. However, since you ask for my advice, I should draw your attention to the following:

[*insert any mitigating information on behalf of the contractor*]

You may or may not wish to take this information into account when making your decision.

Yours faithfully,

he may be aware of the possible consequences of deduction (see letters 5 and 6).

It is not part of your duty (except in certain circumstances under the ACA82 form, see section 1.4.2) to inform the employer how much money he is entitled to deduct. In practice, however, he will probably ask and you can hardly refuse the information. It is best, however, to let him do the final calculation (see letter 18).

The decision to deduct is entirely for the employer. If he asks your advice, you must phrase your reply carefully (see letter 19). If you sincerely think that, despite the contract provisions, it would be unfair to deduct liquidated damages, it does no harm to indicate your view provided you make it quite clear that the final decision rests with the employer (see letter 20).

9 Architect's certificates

9.1
Timing
You are responsible for issuing a number of certificates during the contract (see tables 7, 8, 9 and 10). It is of critical importance that you issue all certificates at the right time. Delay in certification can give rise to claims and leave you facing an action for negligence from the employer. The right time to issue a certificate is either:

● as soon as it falls due; *or*
● as soon as you are able to issue it in accordance with the contract.

9.2
Content
The certificate must contain all necessary information but no more. Where standard certification forms are available for various purposes, they should be used in preference to a letter because they act as a checklist for the information to be included. If you decide to use a letter for the purpose, you must include the words 'This is to certify' or 'I certify' so that the purpose is clear. It is also a good idea to head the letter: 'Certificate of . . .'

If the certificate is one of a series, it is usual to number it in order of issue. It must be dated and signed by a registered architect (not just the office stamp) or, in the case of the GC/Works/1 form, the superintending officer, who might not be an architect.

Remember to issue the certificate to the correct party, as indicated in the contract, and send copies to all relevant persons.

9.3
Dangers
Do not attempt to amend a certificate. Once issued, it is a contractual document and cannot be altered unless the mistake is obvious and does not alter the effect of the certificate. Such a mistake

Table 7 Certificates to be issued by the architect under JCT80

Clause	Certificate
17.1	Practical completion
17.4	Making good of defects
18.1.1	Partial possession
18.1.3	Making good of partial possession
22A.4.2	Payment of insurance money
24.1	Contractor's failure to complete on due date
27.4.4	Expenses incurred by employer on determination of contract by employer
30.1.1.1	Interim certificates
30.7	Interim certificate including finally adjusted sub-contract sums
30.8	Final certificate
35.15.1 35.16	Delay by nominated sub-contractor

Table 8 Certificates to be issued by the architect under ACA82

Clause	Certificate
1.7	Payment of fees by contractor according to clause 1.7
6.4 (alt 2)	Adjustment to contract sum in respect of restoration, repair and/or removal
11.2	Works not fit and ready for taking-over
12.1	Taking-over certificate
12.3	Adjustment to contract sum in respect of clause 12.2
13.3	Reduction of the liquidated and ascertained damages sum
16.2	Interim certificates
17.5	Adjustment to contract sum in respect of instructions and damage, loss and/or expense
19.1	Final certificate
22.1	Damage, loss and/or expense incurred by employer after termination
22.2	Amount due to contractor after termination

Table 9   Certificates to be issued by the architect (superintending officer) under GC/Works/1

Clause	Certificate
24	Vouchers for daywork
28A(1)	Satisfactory completion of part of the works
40(3)	Certification of advances on account
40(6)(b)	Amount not paid to nominated sub-contractor or supplier
42(1)	Certificate of payment under clauses 40 and 41 Satisfactory completion of works Works in a satisfactory state at the end of the maintenance period
46(1)(e)	Cost of completion after determination

Table 10   Certificates to be issued by the architect under the Agreement for Minor Building Works

Clause	Certificate
2.4	Practical completion
2.5	Making good of defects
4.2	Progress payments
4.3	Penultimate certificate
4.4	Final certificate

might be an incorrect name or an arithmetical error.

Certificates are sometimes lost in the post or delayed. All certificates should be sent by recorded delivery. Financial certificates are best delivered by hand if possible and a receipt obtained.

10 Employer's decisions

10.1
General

There are some decisions, in the course of any contract, which can be taken only by the employer. This is usually because the contract:

- reserves certain decisions to the employer; and/or
- restricts the authority of the architect.

In addition, there are some situations which are not specifically envisaged by the contract. Two types of claim fall under this heading:

- ex-contractual claims
- *ex-gratia* claims.

They were discussed in section 1.3. If you receive a claim that you consider to be ex-contractual or *ex-gratia* or if such a claim forms part of a larger claim with which you are dealing, it must be passed to the employer without delay. Do not be tempted to pass any opinion regarding the virtue of the claim unless asked (see letter 21). On receipt of your letter, the employer will almost certainly request your advice. You must advise him to see his solicitor, although you should be ready to accompany him, because ex-contractual claims usually result from a breach of contract and may have to be settled in court or arbitration (see letter 22). If the claim is purely *ex-gratia* and admitted to be such by the contractor, it is entirely a matter for the employer. Legalities do not enter into it and your view is irrelevant. Let the contractor know what you are doing (see letter 23). Whether to deduct liquidated damages is always the employer's decision.

10.2
JCT80

The employer has no part to play in estimating extensions of time and you must firmly resist any

21   Letter from architect to employer regarding ex-contractual or
ex-gratia claims
This letter is not suitable for use with the Agreement for Minor Building Works

Dear Sir,

I enclose a claim dated [*insert date*] sent to me
by the main contractor.

The contract makes provision for me to deal with
financial claims of specific kinds. This claim
is/Parts of this claim are [*omit as appropriate*]
outside those provisions and, therefore, I have no
authority to ascertain either validity or payment.

Although the decision is one for you alone, I am
always ready to assist with further information
and advice if you require it.

Yours faithfully,

22   Letter from architect to employer if advice requested on the matter of ex-contractual or *ex-gratia* claims

Dear Sir,

Thank you for your letter/I refer to your telephone call [*omit as appropriate*] of the [*insert date*].

[*Add the following if the claim is ex-contractual*] The best advice I can give is that you make an appointment to see your solicitor to discuss the position. I will come with you to provide whatever additional information he may need. Do not simply reject the claim at this stage, unless your solicitor so advises, because the contractor may decide to pursue the matter through the courts or in arbitration.

[*Add the following if the claim is purely ex-gratia*] The contractor's claim appears to be purely ex-gratia. In other words, it has no legal basis. The contractor is simply asking you to consider authorising payment because he thinks he has suffered hardship. I will give you any further information you may require to enable you to reach a decision. Even if you come to the conclusion that the contractor has indeed suffered hardship, you are under no obligation to pay anything.

Yours faithfully,

23   Letter from architect to contractor if ex-contractual or *ex-gratia* claim submitted

Dear Sir,

I have received your claim dated [*insert date*] which has been passed to the employer for his decision because it is outside the limits of my authority to decide claims of this nature.

Yours faithfully,

attempts on his part to influence your award. The decision of the employer and the agreement of the contractor are required if acceleration measures are to be used.

Claims for loss and/or expense under the contract provisions are matters for you alone.

10.3
ACA82

The employer must not be concerned with the estimation of extension of time; nor is he contractually required to agree to acceleration measures. However, your own agreement with the employer will require you to obtain the employer's approval to acceleration measures unless you intend to pay for them yourself!

Claims for damage, loss and/or expense under the contract provisions are matters for you alone.

10.4
GC/Works/1

Extensions of time under this contract are reserved to the authority to decide although undoubtedly you will be expected to prepare a full report and recommendations for the authority's consideration. Acceleration measures are also a matter for the authority.

Claims for expense under the contract provisions are matters for your decision but, in the context of this particular contract, you should submit your findings to the authority for approval.

10.5
Agreement for Minor
Building Works

Extensions of time under this contract are entirely your concern.

Claims for loss and/or expense are not provided for. You must, therefore, refer each claim to the employer for his decision and tell the contractor what you are doing (see letter 23). The general procedure outlined in section 10.1 applies. You must make clear to the employer that he cannot simply shrug off a claim because there is no provision in the contract. If the claim has substance, the contractor can pursue the matter at common law. It is likely to be less expensive and wearing to settle it by correspondence and agreement (see letter 24). The employer will be grateful if you let him have a report, with the

24   Letter from architect to employer if contractor submits claim for
loss and/or expense
This letter is suitable only for use with the Agreement for Minor Building Works

Dear Sir,

I enclose a claim for loss and/or expense dated
[*insert date*] which has been submitted by the main
contractor. The contract provisions do not allow
me to ascertain validity or payment for such
claims.

Do not reject the claim out of hand, the
contractor may decide to pursue it through the
courts, which could be expensive. If you wish to
discuss the claim with me, please telephone and a
time can be arranged. It may be possible to
settle the matter quite easily with the contractor.
At our meeting, we could discuss whether it is
necessary to obtain legal advice on this
particular point.

Yours faithfully,

contractor's claim, setting out the facts.

If the contractor submits a large number of small financial claims throughout the contract period, the employer may consider extending your authority as agent to deal with them for him. It is an onerous task and, if you agree, you should make sure you obtain his individual agreement to each decision before you put it into effect. An increase in your fee is also indicated.